LOW-VOLTAGE CMOS RF FREQUENCY SYNTHESIZERS

A frequency synthesizer is one of the most critical building blocks in any wireless transceiver system. Its design is getting more and more challenging as the demand for low-voltage low-power high-frequency wireless systems continuously grows. As the supply voltage is decreased, many existing design techniques are no longer applicable. This book provides the reader with architectures and design techniques that enable CMOS frequency synthesizers to operate at low supply voltages, at high frequencies with good phase noise and with low power consumption. In addition to updating the reader on many of these techniques in depth, this book will also introduce useful guidelines and step-by-step procedures on behaviour simulations of frequency synthesizers. Finally, three successfully demonstrated CMOS synthesizer prototypes with detailed design consideration and description will be presented to illustrate potential applications of the architectures and design techniques described. The book is intended for engineers, managers and researchers who are working or interested in radio-frequency integrated-circuit design for wireless applications.

HOWARD C. LUONG is an associate professor at the Hong Kong University of Science and Technology. He has published over 50 conference and journal papers on low-voltage wireless transceivers and synthesizer designs. He holds two US patents and has five pending. He is a co-author of *Design of Low-Voltage CMOS Switched-Opamp Switched-Capacitor Systems* published by Kluwer Academic Publishers in 2003.

GERRY C. T. LEUNG was a research assistant at the Hong Kong University of Science and Technology. He is currently working as an RFIC designer at Innovative Systems Corporated, Hong Kong.

T0185908

LOW-VOLTAGE CMOS RF FREQUENCY SYNTHESIZERS

HOWARD CAM LUONG AND GERRY CHI TAK LEUNG

Department of Electrical and Electronic Engineering
Hong Kong University of Science and Technology

CAMBRIDGE UNIVERSITY PRESS
Cambridge, New York, Melbourne, Madrid, Cape Town, Singapore,
São Paulo, Delhi, Dubai, Tokyo, Mexico City

Cambridge University Press
The Edinburgh Building, Cambridge CB2 8RU, UK

Published in the United States of America by Cambridge University Press, New York

www.cambridge.org
Information on this title: www.cambridge.org/9780521153492

First published 2004
First paperback printing 2010

A catalogue record for this publication is available from the British Library

Library of Congress Cataloguing in Publication data
Luong, Howard C. (Howard Cam)
Low-voltage RF CMOS frequency synthesizers/Howard Cam Luong and Gerry Chi Tak Leung.
p.   cm.
Includes bibliographical references and index.
ISBN 0 521 83777 4
1. Frequency synthesizers – Design and construction.   2. Metal oxide semiconductors,
Complementary – Design and construction.   I. Leung, Gerry Chi Tak, 1979–   II. Title.
TK7872.F73L86   2004
621.3815′486 – dc22   2004045182

ISBN 978-0-521-83777-4 Hardback
ISBN 978-0-521-15349-2 Paperback

Contents

Figures

Tables

Preface

A frequency synthesizer is one of the most critical building blocks in any integrated wireless transceiver system. Its design is getting more and more challenging as the demand for low-voltage low-power high-frequency wireless systems is continuously increased. At the same time, CMOS processes have advanced and been shown to be more and more attractive due to their potential in achieving systems with the highest integration level and the lowest cost. On the other hand, as the supply voltage is lowered, many existing design techniques for integrated frequency synthesizers are no longer applicable. However, it is still desirable to design RF frequency synthesizers at low supply voltages not only because of the device reliability due to the technology scaling but also because of the integration and compatibility with digital circuits.

There are currently only a few books available on integrated RF CMOS frequency synthesizers. The most comprehensive book on integrated CMOS frequency synthesizers available today is entitled *Wireless CMOS Frequency Synthesizer Design* by Craninckx and Steyaert (1998). More recently, another book entitled *Multi-GHz Frequency Synthesis and Division* by Rategh and Lee was also published in 2001. While the two books are still quite useful, they focus only on advanced design techniques of some selected building blocks, including voltage-controlled oscillators, dividers, and synthesizers, with emphasis only on a particular architecture. There exist many new synthesizer architectures and design techniques that are not covered in detail.

This book is intended to supplement the two books with more comprehensive and in-depth descriptions of building blocks and synthesizer systems, in particular for applications with low supply voltages and high frequencies. Special emphasis is placed on consideration, comparison, trade-offs, and optimization for different design choices. In addition, useful guidelines and step-by-step procedures on behavior simulations of frequency synthesizers will be introduced. Finally, several chip prototypes that were successfully designed and demonstrated at low supply voltages will be described in detail to illustrate potential applications of the architectures and the design techniques presented.

The first prototype demonstrates a fully integrated dual-loop synthesizer for 900 MHz GSM transceivers in a standard 0.5 μm CMOS process with a supply voltage of 2 V and threshold voltages around 1 V. For fair comparison, the second chip prototype employs fractional-N synthesizer architecture with sigma–delta modulation for the same GSM system using the same 0.5 μm CMOS process but with a supply voltage of 1.5 V. The third prototype focuses on a monolithic 1 V, 5.2 GHz integer-N synthesizer for the WLAN 802.11a transceiver system in a 0.18 μm CMOS process with threshold voltages around 0.5 V. While the first two GSM synthesizers need to cover a frequency band of 25 MHz with a channel spacing of 200 kHz around 900 MHz, the third WLAN synthesizer requires a frequency band of 200 MHz with a channel spacing of 20 MHz around 5.2 GHz.

Acknowledgements

It is our great pleasure to have this opportunity to acknowledge and to express our gratitude to many people who have directly or indirectly contributed to this work.

We would like to thank William Shing-Tak Yan and Bob Chi-Wa Lo for their great work and their contribution on dual-loop and fractional-N frequency synthesizers, which play an important part of this book. Our special thanks go to Fred Kwok for his enthusiastic and indispensable support in preparing the test equipment and keeping it in good condition.

We are whole-heartedly grateful to many other students in the analog research laboratory at the Electronic and Electronic Engineering Department (EEE) of the Hong Kong University of Science and Technology (HKUST): namely Toby Kan, Joseph Wong, Vincent Cheung, Chunbing Guo, Lincoln Leung, Ka-Chun Kwok, Gary Wong, Kenneth Ng, Ka-Wai Ho, Thomas Choi, Issac Hsu, David Leung, and Bunny Mak. Without their research efforts and without many fruitful discussions with them, it would have been impossible for us to acquire a good understanding of the topic.

The technical support and assistance by many technical officers in the EEE department at HKUST, in particular Siu-Fai Luk, Franky Leung, Kenny Pang, John Law, and Jacob Lai, are greatly appreciated.

Lastly, we are indebted to our family members for their constant love, support, encouragement, and patience throughout the projects and during the writing of this book.

1

Introduction

1.1. Motivation

Modern transceivers for wireless communication consist of many building blocks, including low-noise amplifiers, mixers, frequency synthesizers, filters, variable-gain amplifiers, power amplifiers, and even digital signal processing (DSP) chips. Each of these building blocks has a different specification, imposes different constraints, and requires different design considerations and optimization. As a result, wireless transceivers have been exclusively implemented using hybrid technologies, mainly GaAs for low noise and high speed, bipolar for high power, passive devices for high selectivity and CMOS for DSP at the baseband. While taking advantage of the best in each technology, this hybrid combination unfortunately requires multi-chip modules and off-chip components, which not only are costly and bulky, but also consume a lot of power.

However, recent development and advance scaling of deep-submicron CMOS technologies have made it more feasible and more promising to implement a single-chip CMOS wireless transceiver. This single-chip integration is particularly attractive for its potential in achieving the highest possible level of integration and the best performance in terms of cost, size, weight, and power consumption.

Among the many design issues and considerations in single-chip CMOS integration is the aggressive scaling of the channel length. According to the Semiconductor Industry Association's roadmap in November 2001, the channel length will be scaled to be as small as 65 nm in 2007, as illustrated in Fig. 1.1. Such a small channel length is necessary to increase operation frequency and to reduce chip area but, at the same time, inevitably requires low supply voltages to avoid oxide breakdown. Even though the threshold voltages of CMOS devices would also be reduced, they are not scaled in the same proportion as the channel length and the supply voltage because of the leakage current in digital circuits. Moreover, the dynamic power of digital circuits is quadratically proportional to their supply voltages. Consequently,

Introduction

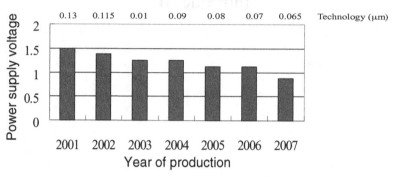

Fig. 1.1 Semiconductor Industry Association's roadmap

the supply voltages for digital circuits have been continuously and aggressively lowered to save power and to extend battery life. To maintain compatibility with the digital parts in wireless transceiver systems, it is necessary that the analog part is designed to operate at the same low supply voltage. Unfortunately, unlike for the digital counterpart, such a low-voltage constraint unavoidably and significantly degrades the performance of analog circuits and, thus, of the whole system, unless novel analog circuit design techniques are developed to compensate and to maintain the same performance.

On-chip voltage multipliers and DC-to-DC boost converters have been proposed to increase the low supply voltage for digital circuits to a higher voltage level in order to power analog circuits and maintain the same performance (Dehng *et al.*, 2000). However, such voltage converters occupy a large chip area, consume extra power, and contribute switching noise to the system. This is the main reason why novel analog design techniques that are suitable for implementation with low supply voltages have become more and more attractive and popular.

It is worth emphasizing that for analog integrated circuits, low-voltage designs do not necessarily result in low power. As a matter of fact, as the supply voltage is lowered, the current consumption typically needs to be increased to maintain the same performance, which in the end can result in even larger total power consumption. In other words, it is always desirable to achieve low power by going to a low supply voltage, but low power should not be the main reason for low-voltage designs. Device reliability due to technology scaling and compatibility with digital circuits in mixed-signal systems should be the main considerations.

One of the greatest challenges to integrating low-voltage single-chip CMOS transceiver systems is to design fully integrated frequency synthesizers for frequency translation and channel selection. First of all, due to very narrow channel spacing, the output signal of the synthesizers needs to be extremely stable and accurate. As a consequence, the phase noise and the spurious performance should be very good. Moreover, the synthesizer's output needs to oscillate at a very high frequency with a sufficiently wide frequency tuning range to cover the whole frequency band and, at the same time, to compensate for any frequency deviation due to process variation. Finally, all these stringent specifications need to be met with limited power consumption and small chip area.

There are several different types of synthesizer architecture, including direct analog synthesis, direct digital synthesizers (DDS) and phase-locked loop (PLL)-based synthesizers (Goldberg, 1996; Yamagishi *et al.*, 1998). Among them, the PLL-based synthesizer is generally most suitable for radio-frequency applications and in general consumes less power consumption with a smaller chip area.

The focus of this book is on design techniques for low-voltage RF CMOS PLL-based frequency synthesizers for wireless transceiver systems. Roughly, low voltage refers to any supply voltage of around 40% to 60% of the maximum allowed supply voltage for a particular process while not exceeding two or three times the threshold voltages of the devices. As examples, in a 0.35 μm CMOS process with a maximum supply of 3.3 V and a threshold voltage of around 0.75 V, a design with a supply voltage of around 1.5 V to 2.0 V is considered a low-voltage design. Similarly, the emphasis is on 1 V designs in 0.18 μm CMOS processes with a maximum supply voltage of 1.8 V and a threshold voltage of around 0.5 V.

1.2. Book organization

The organization of the book is as follows. In Chapter 2 PLL fundamentals and different PLL-based synthesizers will be reviewed together with some brief and qualitative comparison. Chapter 3 will discuss the design issues of the required building blocks and components of PLL synthesizers, including voltage-controlled oscillators, dividers, programmable prescalers, phase frequency detectors, charge pumps, loop-filters, on-chip inductors, varactors, and switched-capacitors arrays. In Chapter 4, guidelines and step-by-step procedures to perform behavioral modeling and simulations of PLL will be presented. Chapter 5 addresses, and elaborates on, special design issues and techniques suitable for high-frequency low-power integrated synthesizers at low supply voltages. As a demonstration of potential applications of the system architectures and design techniques discussed, Chapters 6, 7 and 8 will present detailed design considerations, practical issues, and the successful implementation of several state-of-the-art frequency synthesizers: namely a 2 V dual-loop frequency synthesizer, a 1.5 V 900 MHz fractional-N synthesizer with sigma–delta modulation, and a 1 V 5.2 GHz integer-N synthesizer in 0.5 μm, 0.5 μm and 0.18 μm CMOS processes, respectively. Finally, conclusions are drawn in Chapter 9.

2

Synthesizer fundamentals

2.1. Introduction

Nowadays, many integrated circuits are operated in the multi-gigahertz range to increase their processing power and data bandwidth. High-speed clock generation is necessary for both RF systems and microprocessor systems. For high-frequency synchronous systems, the clock fluctuation needs to be minimized to prevent race conditions, to shorten the setup time and hold time requirements, and to increase the maximum possible operating speed of clocked systems.

Local oscillators (LOs), key elements in transceivers, are required to down-convert or up-convert RF signals while minimizing degradation of the signal-to-noise ratio (SNR). The LO signal is expected to be an ideal tone, which should be stable and clean and appear as a sharp impulse. Unfortunately, in practical situations, intrinsic noise from devices and noise from the surrounding environment make the LO signal fluctuate. As a result, the LO signal appears with sideband noise as a skirt centered around the impulse in the frequency domain. For wireless applications, this noise performance affects the SNR and is characterized by measuring the phase noise, which is defined as the ratio of the power of the signal at the desired frequency to the power of the signal at an offset frequency. For clocked system applications, jitter is normally used to characterize timing uncertainty of a clock signal in the time domain, which is defined as the deviation of the zero-crossing points from the ideal waveform.

In order to generate a high frequency and stable clock signal, frequency synthesis is necessary. Among many choices, frequency synthesizers using a phase-locked loop (PLL) are the most popular, in particular for high-frequency and low-power signal generation. Basically, a PLL-based synthesizer is a feedback system used to generate a stable clock signal based on a reference signal. The performance of the synthesizer depends heavily on the purity of the reference signal. As a

5

result, crystals are commonly used to generate the reference frequency in most PLL systems because of their excellent purity and stability.

The first sections of this chapter review noise sources, which contribute to degradation of phase noise, and fundamental concepts. The trade-offs between design parameters in PLL-based synthesizers are also discussed. The second part of the chapter will briefly describe and compare different architectures to implement high-frequency PLL synthesizers.

2.2. Timing jitter

Timing jitter is a statistically measured parameter with a zero-mean Gaussian distribution and is used to characterize the noise performance of clock signals in the time domain. Deviations of the zero-crossing of the rising clock edge or falling clock edge from their ideal positions are quantified by jitter. Long-term jitter, or absolute jitter, $\sigma_{\Delta T}$, can be used to measure the jitter performance of an oscillator and it is expressed as

$$\sigma_{\Delta T} = \sum_{n=1}^{N} (T_n - \bar{T}), \tag{2.1}$$

where T_n is the period of the oscillator output at the nth cycle and $\bar{T}$ is the mean oscillation period (Herzel and Rajavi, 1999). Owing to the existence of flicker noise in CMOS oscillators, the standard deviation of jitter of free running Voltage-controlled Oscillators (VCOs) is proportional to the square root of the measured time before $t_{1/f}$, and directly proportional to the measured time after $t_{1/f}$, where $t_{1/f}$ is the measured time associated with the $1/f$ noise of devices in oscillators. The timing jitter of a free-running oscillator, $\sigma(t)$, is described in Equation (2.2) (Hajimiri, Limotyrakis and Lee, 1999):

$$\sigma(t) = c\sqrt{t} + kt, \tag{2.2}$$

where c and k are constants with and without $1/f$ noise, respectively.

Large jitter will degrade the accuracy of clocked systems for synchronized operations and cause intercommunication errors between systems. A high-performance clocked system requires stringent clock stability. An unstable clock may limit the maximum possible speed of clocked systems and may even cause incorrect data sampling. Thermal noise in devices can cause phase noise and amplitude noise as depicted in Fig. 2.1.

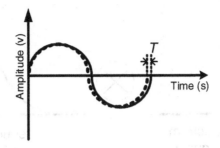

Fig. 2.1 Clock with timing jitter

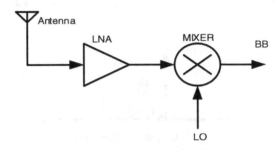

Fig. 2.2 Receiver with LO signal

2.3. Phase noise

Phase noise is defined as the ratio between the total carrier power and the noise power at a frequency offset from the carrier, Δf, which is shown in Equation (2.3).

$$L(\Delta f) = 10 \log \left(\frac{\text{power in 1 Hz bandwidth at } f_0 + \Delta f \text{ frequency offset from carrier}}{\text{total carrier power}} \right),$$

(2.3)

where $L(\Delta f)$ is the phase noise in units of decibels per hertz (dBc/Hz), and f_0 is the center frequency of the oscillator.

The higher the required SNR of an RF system is, the better the phase noise for an oscillator is expected. For the receiver in Fig. 2.2, if the LO signal exhibits a non-ideal phase noise as shown in Fig. 2.3, both the desired RF signal and the interference can be simultaneously mixed down by the LO signal. After mixing the interferer can fall directly in the same band as that of the desired signal. Consequently, the SNR is unavoidably degraded. This effect is generally referred to as reciprocal mixing. Since the interference signals are typically much larger than the desired RF signal, the SNR may be unacceptably small unless the phase noise of the LO signal is sufficiently low.

Amplitude noise and phase noise exist in any oscillator. Owing to amplitude limitation in the practical oscillator, phase noise is more severe than amplitude noise (Hajimiri, Limotyrakis and Lee, 1999). Any phase error that occurs in earlier

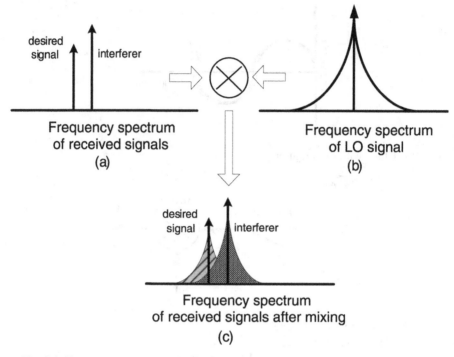

Fig. 2.3 Frequency spectrum of (a) received signal, (b) LO signal with phase noise, and (c) received signal suffering from SNR degradation after mixing

transitions of the oscillator will be accumulated over time, and the oscillator itself cannot recover the phase error. Therefore, the design specification of phase noise for an oscillator is stringent, especially in CMOS technology.

An oscillator's phase error can be shown to increase with measured time when the oscillator is free running. This also leads to a frequency shift. Hence, a control mechanism is required to lock the system and keep it stable (as will be mentioned in detail in Section 2.4).

In fact, phase noise and jitter are related to each other. However, phase noise represented in the frequency domain can give a more detailed picture of how noise contributes at different frequency offsets from the carrier frequency.

Phase noise has been shown to have a Lorentzian spectrum and is depicted as (Poore, 2001)

$$L(\Delta f) \propto \frac{f_o^2}{\left(\alpha \pi f_o^2\right)^2 + \Delta f^2}, \tag{2.4}$$

where f_o is the oscillating frequency and Δf is the frequency offset from the carrier. The noise spectrum falls at $-20\,\text{dB}$ per decade. However, it does not include flicker noise (or $1/f$ noise) and white noise. Leeson's equation shown in Equation (2.5)

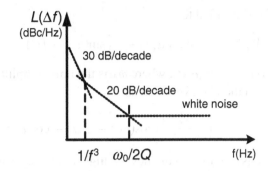

Fig. 2.4 Noise spectrum of typical oscillators

describes the detail of the phase noise spectrum in practical oscillators (Leeson, 1966; Lee, 2000):

$$L(\Delta f) = 10 \log \left[\frac{2FKT}{P_0} \left(1 + \frac{f_0^2}{4Q^2 \Delta f^2} \right) \left(1 + \frac{\Delta f_{1/f^3}}{\Delta f} \right) \right], \qquad (2.5)$$

where:

F	is the excess noise factor;
K	is the Boltzmann constant with a value of 1.38×10^{-23} J/K;
T	is the absolute temperature;
P_0	is the power of the carrier signal;
f_0	is the carrier frequency;
Q	is the quality factor in the LC tank;
Δf	is the offset frequency from the carrier frequency;
$\Delta f_{1/f^3}$	is the corner frequency of $1/f$ noise.

Thus, Leeson's model includes flicker noise and the white noise floor. The noise spectrum is depicted in Fig. 2.4.

Lesson's model indicates that typical phase noise falls at -30 dB per decade before the $1/f^3$ frequency corner. This phenomenon occurs because $1/f$ noise modulates the transconductance of the transistors in the oscillator. Compared with bipolar transistor, CMOS devices exhibit a wider $1/f^3$ frequency region because of their higher $1/f$ frequency corners (Razavi, 1996b). Beyond the $1/f^3$ frequency region, the phase noise drops to -20 dB per decade and finally levels off as white noise becomes dominant. The noise floor is mainly contributed by the thermal noise of the devices and results in the phase noise being equal to $2FKT/P_0$.

In the time domain, the oscillating signal $V_{vco}(t)$ with timing jitter can be expressed as

$$V_{vco}(t) = V_o \cos[\omega_o t + \phi(t)], \qquad (2.6)$$

where $\phi(t)$ is the phase noise of the oscillator, and V_o is the signal amplitude.

The output can be simplified to

$$V_{vco}(t) = V_o \cos(\omega_o t) - V_o \sin(\omega_o t) \cdot \phi(t). \tag{2.7}$$

If $\phi(t)$ is assumed to be $\phi_A \sin(\omega_\phi t)$, where ϕ_A is the noise amplitude at frequency ω_ϕ, Equation (2.7) can then be expressed as

$$V_{vco}(t) = V_o \cos(\omega_o t) - \frac{V_o \phi_A}{2}[\cos(\omega_o t - \omega_\phi t) - \cos(\omega_o t + \omega_\phi t)]. \tag{2.8}$$

It shows that any noise at low frequency or high frequency can be modulated and appear as the sideband of the carrier in the frequency domain. Thus, noise consideration in oscillator designs should not be limited only to frequencies around the fundamental oscillation frequency.

To design receivers with a required SNR, the phase noise of the oscillator should satisfy the condition

$$L(\Delta\omega) < S_{RF} - S_{block} - 10\log_{10}(\text{BW}) - \text{SNR} \tag{2.9}$$

where:

    S_{RF} is the desired RF signal power;
    S_{block} is the blocking signal power;
    BW   is the channel bandwidth of the desired RF signal;
    SNR is the signal-to-noise ratio for a system to achieve the required bit-error rate.

From the system point of view, the maximum power of the blocking signals and thus the phase noise requirement are different at different frequency offsets from the carrier. As a result, the phase noise requirement should be based on the worst case consideration.

2.4. Phase-locked loop

As mentioned in Section 2.2, an oscillator by itself cannot recover the phase error and the frequency shift. However, as long as the phase of the oscillator can be controlled and locked, the frequency can also be locked. This is because the change of the phase with respect to time is equivalent to its frequency. Consequently, a phase-locked loop (PLL) is typically used to lock both the phase and the frequency of the oscillator to provide a stable output signal.

A PLL is a negative feedback system, and a general block diagram of such a system is shown in Fig. 2.5. Basically, a PLL system consists of an input phase detector (PD), a charge pump, a loop filter, a voltage-controlled oscillator (VCO), and dividers. Dividers N and M are optional but may be required depending on the desired ratio between the PLL's output frequency and the reference clock frequency.

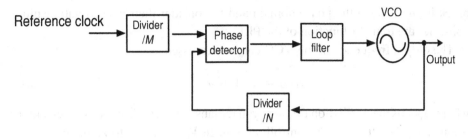

Fig. 2.5 Block diagram of a typical phase-locked loop

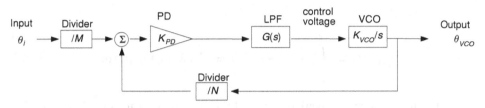

Fig. 2.6 Linear model of PLL system

By choosing different division ratios, the resultant output frequency of the VCO, f_{VCO} can be locked at

$$f_{VCO} = f_{ref}\frac{N}{M}. \qquad (2.10)$$

The phase detector acts as a 'phase-error amplifier' to sense the phase difference between the output signals of the dividers, and to generate a voltage proportional to the phase difference, which in turn is applied to the charge pump and the loop filter to control the VCO output frequency. Accordingly, the output frequency of the divider N is varied and is compared with the output frequency of the divider M by the phase detector. With the loop closed, the phase of the VCO signal is continuously being detected and adjusted. Eventually, the frequency of the VCO is equal to the frequency of the reference clock multiplied by N/M.

The reference clock can modulate the VCO and appear as spurious tones in the frequency spectrum. To reduce the tones, a low-pass filter (LPF) can be used between the PD and the VCO to filter out the high-frequency components.

The linear model shown in Fig. 2.6 can be used to analyze the loop behavior of a typical PLL in a locked condition. This is useful for analyzing both the stability of the loop and the noise contributed by individual building blocks to the output of the PLL. To achieve the locked condition, the frequency and phase acquisition processes need to take place as verified by the behavioral simulation shown in Section 5.5. Since the phase is the parameter of interest rather than the oscillation frequency, it is necessary to represent the input and the output of each building block in the PLL in the phase domain and to consider their phase relationship to construct the transfer function of the PLL.

The output frequency of the VCO can be expressed as

$$\omega_{VCO} = \omega_0 + K_{VCO} V_{ctrl}, \tag{2.11}$$

where ω_{VCO} is the oscillation frequency in radians per second, ω_0 is the free-running frequency when the VCO's controlling voltage V_{ctrl} is equal to zero, and K_{VCO} is the gain of the VCO. The excess phase of the VCO, which is defined as the extra phase shift introduced by the control voltage multiplied by the VCO gain, is given by

$$\theta_{VCO}(t) = \int_{-\infty}^{t} (\omega_{VCO} - \omega_0)\, dt = \int_{-\infty}^{t} (K_{VCO} V_{ctrl}(t))\, dt \tag{2.12}$$

as the frequency is the derivative of the phase with respect to time. The transfer function of the VCO in the s domain becomes

$$\frac{\theta_{VCO}(s)}{V_{ctrl}(s)} = \frac{K_{VCO}}{s}. \tag{2.13}$$

Therefore, the phase is changed after the frequency is changed and the integration is performed. To understand the locking behavior, consider the situation when the VCO is oscillating at a frequency larger than the expected frequency. In this case, the rate of accumulated phase error is positively increasing, and the negative feedback loop will decrease the control voltage of the VCO to pull down its output frequency, which in turn results in a reduction to zero in the rate of accumulated phase error. The system itself eventually achieves the desired output frequency and a zero phase error at the inputs of the PD. The feedback loop will stop adjusting the control voltage until a phase error is detected again by the PD.

Similarly, the frequency divider performs a frequency division with division ratio N, and so the frequency of the input signal $f_{div\text{-}in}$ and the output signal $f_{div\text{-}out}$ can be related as

$$2\pi f_{div\text{-}out} = \frac{2\pi f_{div\text{-}in}}{N}$$

$$\int_{-\infty}^{t} \omega_{div\text{-}out}\, dt = \frac{\int_{-\infty}^{t} \omega_{div\text{-}in}\, dt}{N}. \tag{2.14}$$

After integration, the phase transfer function of the divider becomes

$$\frac{\theta_{div\text{-}out}}{\theta_{div\text{-}in}} = \frac{1}{N}. \tag{2.15}$$

The PD produces a corrected voltage and is filtered by the loop filter to vary the VCO frequency. Thus, the transfer function between the input and the output of the PD is given as

$$V_{PD\text{-}out} = K_{PD}(\theta_1 - \theta_2) = K_{PD}\theta_{PD\text{-}in}$$

$$\frac{V_{PD\text{-}out}}{\theta_{PD\text{-}in}} = K_{PD}. \tag{2.16}$$

By considering the PLL system as a linear feedback system, the transfer function of the PLL can be written as

$$H(s) = \frac{G_1(s)}{1 + G_2(s)} \frac{1}{M}, \tag{2.17}$$

where $G_1(s)$ is the forward gain transfer function and $G_2(s)$ is the total loop gain of the PLL system. The forward gain is

$$G_1(s) = \frac{K_{PD}G(s)K_{VCO}}{s}, \tag{2.18}$$

and the loop gain is

$$G_2(s) = G_1(s)\frac{1}{N} = \frac{K_{PD}G(s)K_{VCO}}{sN}. \tag{2.19}$$

So Equation (2.17) can be written as

$$H(s) = \frac{NK_{PD}G_{LPF}(s)K_{VCO}}{sNM + K_{PD}G_{LPF}(s)K_{VCO}M}. \tag{2.20}$$

If a simple RC low-pass filter is employed in the loop filter, and its transfer function is

$$G_{LPF}(s) = \frac{1}{1 + sRC} = \frac{1}{1 + \dfrac{s}{\omega_{LPF}}}, \tag{2.21}$$

the transfer function of the PLL becomes

$$H(s) = \frac{NK_{PD}K_{VCO}\omega_{LPF}}{s^2NM + s\omega_{LPF}NM + K_{PD}K_{VCO}\omega_{LPF}M}. \tag{2.22}$$

This is, in fact, a standard second-order transfer function, which can be rewritten as

$$H(s) = \frac{\omega_n^2}{s^2 + 2s\xi\omega_n + \omega_n^2} \frac{N}{M}, \tag{2.23}$$

where the neutral frequency ω_n of the system is

$$\omega_n = \sqrt{\frac{K_{PD}K_{VCO}\omega_{LPF}}{N}} \tag{2.24}$$

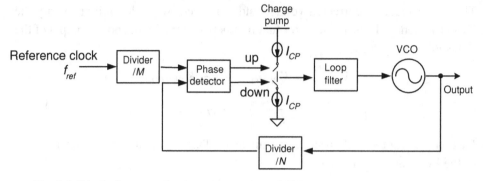

Fig. 2.7 Block diagram of a charge-pump-based PLL

and the damping factor ξ is

$$\xi = \frac{\omega_{LPF}}{2\omega_n} = \frac{1}{2}\sqrt{\frac{\omega_{LPF}N}{K_{PD}K_{VCO}}}. \tag{2.25}$$

Thus, to ensure that the response of the closed-loop system with critical damping optimizes both the bandwidth and the stability, the damping factor ξ should be designed to be around 0.7. Moreover, if ω_n is designed to be larger to achieve a shorter settling time, ω_{LPF} should also be increased proportionally to avoid instability.

2.4.1. Charge-pump-based phase-locked loop (CP-PLL)

The charge-pump-based PLL (CP-PLL) is a popular type of PLL, which employs a charge pump at the output of the phase detector to deliver charge to the loop filter. Such an implementation has the advantage of providing acquisition aid when the loop is out of lock (Egan, 2000). Figure 2.7 shows the block diagram of a CP-PLL.

The phase detector of a CP-PLL produces 'up' and 'down' signals to drive the charge pump to charge or discharge the loop filter so as to generate a control voltage for the VCO. The duty cycles of the 'up' and 'down' signals are based on the phase error of the two PD inputs, which are θ_1 and θ_2, or $\theta_{PD\text{-}in} = \theta_1 - \theta_2$, so the phase transfer function of the PD associated with the charge pump can be written as

$$I_{CP\text{-}out} = I_{CP}\frac{\theta_1 - \theta_2}{2\pi}$$
$$\frac{I_{CP\text{-}out}}{\theta_{PD\text{-}in}} = \frac{I_{CP}}{2\pi}, \tag{2.26}$$

where I_{CP} is the nominal current of the charge pump while the PD is assumed to have linear phase characteristic from -2π to $+2\pi$, as illustrated in Fig. 2.8. If a different PD topology is employed, the gain should be adjusted accordingly.

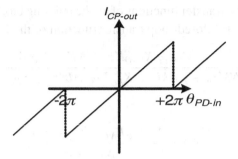

Fig. 2.8 Current-to-phase relation of PD associated with CP

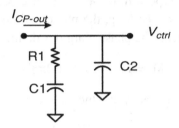

Fig. 2.9 Second-order loop filter

Since a CP-PLL consists of two poles at the origin, one of which is from the VCO and the other is from the integrator in the loop filter, a stabilizing zero is required to prevent the loop from being unstable. A second-order loop filter (shown in Fig. 2.9) is commonly used in a CP-PLL. A resistor R1 is connected in series with a capacitor C1 to provide a stabilizing zero. C2 is connected in parallel to provide filtering capability to reduce ripple at the VCO's control voltage so as to suppress spurious tones at the output of the VCO.

Since the filter converts the current from the charge pump to the VCO input voltage, the transfer function of the loop filter is expressed as

$$\frac{V_{ctrl}}{I_{CP-out}} = \frac{k'}{s}\frac{1+s\tau_z}{1+s\tau_p} = \frac{1+s(R_1C_1)}{s(C_1+C_2)(1+sR_1(C_1\|C_2))}, \tag{2.27}$$

where:

k' is the time constant of integration equal to $1/(C_1+C_2)$;

τ_z is the time constant that provides a stabilizing zero to the loop which is equal to R_1C_1;

τ_p is the time constant of the pole that suppresses the tone of the reference clock and its higher harmonics. The time constant equals $R_1C_1C_2/(C_1+C_2)$.

There are others filter topologies, including passive filters and active filters, which will be discussed in more detail in Chapter 3.

After calculating the transfer functions of all the building blocks, and substituting into Equation (2.20), the closed-loop transfer function of the CP-PLL will be

$$H(s) = \frac{sNK_{PD}K_{VCO}k'\tau_z + NK_{PD}K_{VCO}k'}{s^3NM\tau_p + s^2NM + sK'_{PD}K_{VCO}Mk'\tau_z + K_{PD}K_{VCO}Mk'}, \qquad (2.28)$$

while the loop gain is

$$G_2(s) = \frac{1}{2\pi} \frac{I_{CP}K_{VCO}k'(1+s\tau_z)}{s^2N(1+s\tau_p)}. \qquad (2.29)$$

As expected, there are two poles located at the origin, and, as such, the PLL is commonly referred to as a type-II third-order CP-PLL (Gardner, 1980).

To analyze the stability of the loop, the phase margin (PM) and the loop gain $G_2(j\omega)$ should be considered. The phase margin is defined as

$$PM = 180° + \angle G_2(j2\pi f_c), \qquad (2.30)$$

where f_c is the cross-over frequency at which the loop gain is equal to 1, that is

$$|G_2(j2\pi f_c)| = 1. \qquad (2.31)$$

To ensure that the loop is stable no matter what kind of topology is adopted, the phase margin of the loop gain should be no less than 45° (Gray and Meyer, 1992). Typically, 60° is a desirable phase margin from a lower noise peaking consideration. The phase margin of the third-order CP-PLL is given in Equation (2.32):

$$\angle G_2(j\omega_c) = 180° + \angle \left| \frac{1}{2\pi} \frac{I_{CP}K_{VCO}k'(1+j\omega_c\tau_z)}{\omega_c^2N(1+j\omega_c\tau_p)} \right|$$
$$\angle G_2(j\omega_c) = 180° + \tan^{-1}(\omega_c\tau_z) - \tan^{-1}(\omega_c\tau_p). \qquad (2.32)$$

2.4.2. Phase noise and jitter of phase-locked loop

The phase noise of a free-running oscillator becomes worse at a smaller frequency offset, as depicted in Fig. 2.4. This phenomenon is also consistent with the fact that the timing jitter of a free-running oscillator increases with respect to the measured time as shown in Equation (2.2). The PLL is a good candidate for the precise control of the phase and the frequency of a high-speed oscillator by using a stable crystal as a reference. Therefore, the phase noise of the oscillator, which is due to phase fluctuation, can be suppressed in the loop under locking. However, the suppression is not identical at all frequency offsets. In fact, the noise suppression is less at smaller frequency offsets except at frequencies very far away from the carrier. Equations (2.33) and (2.34) express the phase-noise power of the third-order

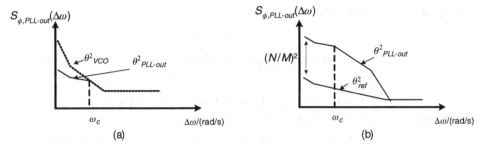

Fig. 2.10 Phase noise in a PLL due to (a) VCO noise, and (b) reference noise

CP-PLL, $\theta^2_{PLL\text{-}out}(s)$, as a function of the phase-noise power of the VCO, $\theta^2_{VCO}(s)$, and that of the reference clock, $\theta^2_{ref}(s)$, respectively.

$$S_{\phi,PLL\text{-}out} = \theta^2_{PLL\text{-}out}(s)$$

$$= \left| \frac{s^3 N\tau_p + s^2 N}{s^3 N\tau_p + s^2 N + sK_{PD}K_{VCO}k'\tau_z + K_{PD}K_{VCO}k'} \right|^2 \theta^2_{VCO}(s), \qquad (2.33)$$

$$S_{\phi,PLL\text{-}out} = \theta^2_{PLL\text{-}out}(s)$$

$$= \left| \frac{sNK_{PD}K_{VCO}k'\tau_z + NK_{PD}K_{VCO}k'}{s^3 NM\tau_p + s^2 NM + sK_{PD}K_{VCO}Mk'\tau_z + K_{PD}K_{VCO}Mk'} \right|^2 \theta^2_{ref}(s). \qquad (2.34)$$

The above equations indicate that the transfer function of the VCO's phase noise to the PLL output is a high-pass response, while that of the reference clock's phase noise exhibits a low-pass characteristic. Thus, the output noise of the PLL depends heavily on the bandwidth of the loop. A large loop bandwidth can reject the timing error of the VCO in a short time and result in higher suppression of the VCO noise to the PLL output. However, it would allow more noise from the input source to the output of the PLL. On the other hand, a small loop bandwidth is able to suppress the noise from the input source as well as the noise coming from the phase–frequency detector and the charge pumps. This phenomenon is illustrated in Fig. 2.10. It should be noted that the phase noise power from the reference clock is amplified by a factor of $(N/M)^2$.

As discussed before, a free-running oscillator exhibits phase variation without being bounded, whereas a closed-loop PLL performs phase detection and adjusts the VCO phase shift. The loop bandwidth of the PLL determines the rate of jitter suppression when VCO is under locking. It has been proved that the jitter of a VCO being locked by a PLL, ΔT_{PLL}, is inversely proportional to the square root of the loop bandwidth, as shown in Equation (2.35) (Hajimiri,

Limotyrakis and Lee, 1999):

$$\Delta T_{PLL} = \sqrt{\frac{1}{2\pi f_c} S_\phi(\Delta f) \frac{\Delta f}{f_o}}, \qquad (2.35)$$

where f_c is the loop bandwidth of the PLL and $S_\phi(\Delta f)$ is the phase noise power at a frequency offset Δf, while f_o is the oscillation frequency.

It is desirable to have a large bandwidth to reduce VCO jitter and to minimize the VCO phase noise, but it is not practical to have a very large loop bandwidth for two reasons. First, the noise from the reference clock, the phase detector, the charge pumps and the dividers may not be attenuated sufficiently and may degrade the overall phase noise. Second, using a continuous-time model to characterize the discrete-time PLL, the loop bandwidth is usually limited to be smaller than one-tenth of the reference frequency to achieve good stability (Razavi, 1998). Thus, in order to account for in-band and out-of-band noise, the jitter can be estimated by integrating the phase noise of the PLL as derived in Hajimiri, Limotyrakis and Lee (1999), and Mansuri and Yang (2002):

$$\Delta T_{PLL} = \frac{2}{\omega_o} \sqrt{\int_0^\infty S_\phi(\Delta f) \, \mathrm{d}\Delta f}. \qquad (2.36)$$

2.4.3. *Spurious tone*

The phase detector in the PLL works at a frequency of f_{ref}/M. The output clocks are switched on and off periodically and drive the successive stage of the charge pump. The mismatch between the pull-up and pull-down currents in the charge pump introduces current injection, ΔI_{CP}, thereby causing ripples at the VCO input.

The ripples modulate the VCO, and the VCO output signal appear with a pair of spurious tones as shown in Fig. 2.11. These kinds of spurious tones are located at frequencies of $f_0 \pm f_{ref}/M$, which are quite close to the carrier frequency. Therefore, as illustrated in Fig. 2.12, as the VCO carrier mixes with the RF signal, the spurious tones may also mix with interference and translate the interference to the same intermediate frequency band of interest. As a result, the SNR is inevitably degraded.

In a third-order CP-PLL, the loop filter capacitor, C_2, (as depicted in Fig. 2.9) is normally used to filter out the reference spurs, and the power of the spurious tone P_{spur} is given by

$$P_{spur} \propto 20 \log \left(\frac{\Delta V_{ctrl}}{\sqrt{2}} \frac{M \cdot K_{VCO}}{f_{ref}} \right) - (n) \cdot 20 \log \left(\frac{f_{ref}}{M \cdot f_p} \right), \qquad (2.37)$$

where ΔV_{ctrl} is the ripple of the VCO's control voltage, f_p is the pole of the loop filter, n is the order of the loop filter, and f_{ref} is the reference frequency of the input clock.

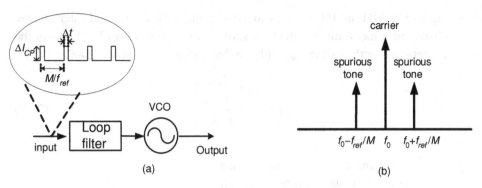

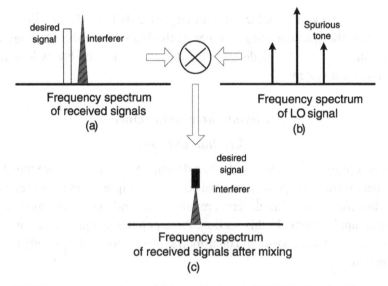

Fig. 2.11 (a) VCO is modulated by the charge injection, (b) VCO output spectrum with spurious tone due to charge injection

Fig. 2.12 Frequency spectrum of (a) received signal, (b) LO signal with spurious tones, and (c) received signal suffering from SNR degradation after mixing

The spur problem can be alleviated by using a smaller loop bandwidth or a loop filter with a higher order to attenuate the spur sufficiently before getting to the input of the VCO. Nevertheless, as a trade-off, both solutions result in a compromise of speed and stability of the PLL design.

2.4.4. Settling time

The settling time is used to define the time required for a synthesizer to switch from one output frequency to another output frequency within a certain frequency accuracy specified by the system requirement. The settling time is a crucial parameter for some systems like Bluetooth[TM], which employ frequency hopping with a hopping rate as high as 1600 hops/s to avoid multi-path fading and interference. Loop bandwidth has a direct effect on settling time. Equation (2.38) shows the relation between settling time t_{lock} and loop bandwidth f_c (Vaucher, 2000):

$$t_{lock} = \frac{\ln\left(\dfrac{f_{step}}{f_{error}}\right)}{f_c \cdot \xi_e(\phi_m)} \tag{2.38}$$

where,

f_{step}	is the amplitude of the frequency jump;
f_{error}	is the maximum frequency error at t_{lock};
ϕ_m	is the phase margin;
$\xi_e(\phi_m)$	is the effective damping coefficient at a specified phase margin.

Faster settling time can be achieved with larger bandwidth. Nevertheless, as discussed earlier, the spurious tone is larger as the loop bandwidth becomes wider. Therefore, there is a severe trade-off when designing the loop bandwidth and the settling time for a system.

2.5. Synthesizer architecture

2.5.1. Introduction

The general criteria to be addressed when designing a synthesizer are the channel spacing specification, the phase noise, the power, the chip area, and complexity. Different architectures can fulfill different requirements and thus can be more suitable for different applications. In this section, three popular synthesizer architectures, namely integer-N, fractional-N and dual-loop synthesizers, will be briefly reviewed and compared.

2.5.2. Integer-N synthesizer

An integer-N frequency synthesizer, as depicted in Fig. 2.13, consists of a programmable divider with an integral division ratio in the feedback path. The phase detector acts as an error detector to sense the phase error between the reference clock signal and the VCO signal after being divided by the programmable divider. The control voltage of the VCO will be dynamically adjusted by the charge pumps and the loop filter until the system is locked. Under the phase-locked condition, the VCO output frequency f_o is equal to the reference frequency f_{ref} multiplied by N.

The output frequency, therefore, can be varied by simply changing the division ratio N. Since N is an integer, it is necessary that the reference frequency is equal to the desired frequency step. For narrow-band systems, the reference frequency should be limited to the channel spacing requirement, while the division ratio N

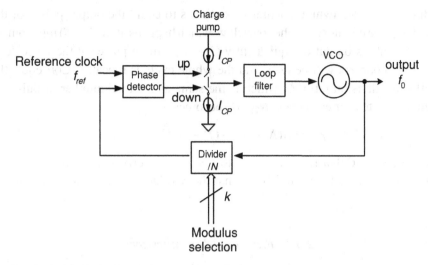

Fig. 2.13 Block diagram of an integer-N frequency synthesizer

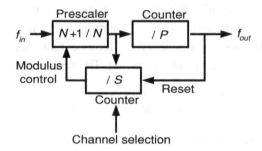

Fig. 2.14 Block diagram of pulse-swallow frequency divider

should be large enough for the VCO to be locked with respect to the reference frequency. Such a large division ratio N results in larger power consumption for the divider and larger phase noise at the synthesizer's output. Moreover, because the loop bandwidth is typically limited to less than one-tenth of the reference frequency for stability, the loop bandwidth and thus the settling time are also limited.

A pulse-swallow frequency divider, as depicted in Fig. 2.14, is the most popular implementation of the programmable prescaler (Akazawa *et al.*, 1983). It consists of two counters, P and S, and a prescaler that is capable of dividing either by $N + 1$ or N according to the modulus control status. Initially, right after the system is reset, the prescaler starts performing divide-by-$(N + 1)$, and both counters simultaneously start counting the output pulses of the prescaler. Counter S keeps counting until its overflow is reached, which occurs after S pulses at the prescaler's output or equivalently $S(N + 1)$ pulses at the prescaler's input. At that instant, the modulus control changes its status, and the prescaler switches from dividing-by-$(N + 1)$ to dividing-by-N. Meanwhile, counter P continues to count the output pulses of the prescaler. Once counter P reaches overflow, which happens after $(P - S)$ more pulses at the prescaler's output or equivalently $(P - S)N$ more pulses at the prescaler's input, both counters will be reset and the whole cycle is repeated. Consequently, the effective division ratio, N, which is the same as the total number of pulses at the input counted for each cycle N_{TOTAL}, becomes

$$N_{TOTAL} = S(N + 1) + (P - S)N = PN + S. \tag{2.39}$$

It should be noted that for the system to operate properly, counter S needs to be reset before counter P, and it follows that the S-value should be smaller than the P-value.

2.5.3. Fractional-N synthesizer

As mentioned in the previous section, integer-N architecture suffers from problems with limited reference frequency and a high division ratio. This architecture is, therefore, not a suitable candidate for use in systems with narrow channel spacing and fast settling time requirements. As a solution, fractional-N synthesizers can provide fractional division ratios so that the reference frequency can be much larger than the frequency step or the channel spacing. The resultant output frequency can be expressed as

$$f_o = N \cdot F^* f_{ref}, \tag{2.40}$$

where N is an integer, and F is a fraction.

As shown in Fig. 2.15, a fractional-N frequency synthesizer contains a dual-modulus programmable divider controlled by an accumulator (Riley *et al.*, 1993). In every cycle, the accumulator's output is incremented by k, which is controlled by the channel-selection input. Once the accumulator overflows, the carry bit changes its state from 0 to 1, and the programmable divider changes its division ratio from N to $N + 1$ accordingly. The resolution of the programmable divider is the same as the total number of cycles that the accumulator repeats and depends on the number of bits designed for the accumulator. For example, if the synthesizer uses a 4-bit accumulator and the modulus is chosen to be 10, the bit stream of the carry output is illustrated by Table 2.1.

It is clear that the division ratio is toggled between N and $N + 1$ according to the carry bit sequence. Hence, the average division ratio N_{AVG} is

$$N_{AVG} = N + \frac{K}{2^k}. \tag{2.41}$$

On the other hand, if a higher reference frequency is used while the frequency step is maintained, the accumulator length k should be increased as well.

Table 2.1 *Output pattern of the 4-bit accumulator with modulus equal 10*

Number of cycle	Content in accumulator	Carry bit	Division ratio
0	0	1	$N + 1$
1	10	0	N
2	4	1	$N + 1$
3	14	0	N
4	8	1	$N + 1$
5	2	1	$N + 1$
6	12	0	N
7	6	1	$N + 1$
8	0	1	$N + 1$
9	10	0	N
10	4	1	$N + 1$
11	14	0	N
12	8	1	$N + 1$
13	2	1	$N + 1$
14	12	0	N
15	6	1	$N + 1$
	average division		$N + 0.625$

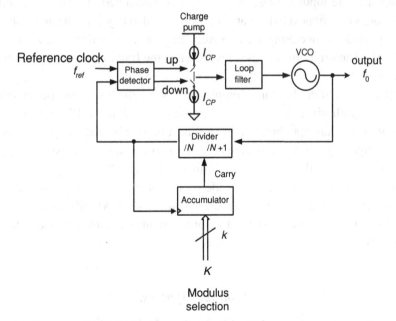

Fig. 2.15 Fractional-N frequency synthesizer

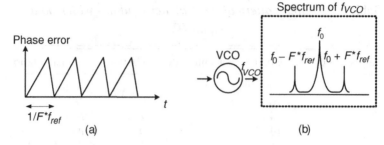

Fig. 2.16 (a) Periodic phase error existing in fractional-N synthesizer, (b) VCO modulated by repetitive signal and its output spectrum

An advantage of using the fractional-N synthesizer is that the reference frequency can be chosen to be much larger than the channel spacing. Thus, the division ratio can be smaller, and the system phase noise can be improved as well. On the other hand, if the reference frequency is larger, the resolution of the accumulator should be increased in order to achieve the same frequency resolution. As a result, the power consumption and the complexity of the accumulator will also need to be increased.

Since the programmable divider is toggled continuously, the phase difference between the reference clock and the divided VCO signal is varied repetitively. It follows that the input voltage of the VCO is modulated up and down and that spurious tones appear next to the carrier as illustrated in Fig. 2.16. It can be observed that the spurious components are relatively large and located at $f_o \pm F \cdot f_{ref}, \ldots,$ $f_o \pm nF \cdot f_{ref}$. The problem becomes severe when the fractional spurs fall within the loop bandwidth.

An effective solution that can be employed to solve the fractional spur problem in fractional-N synthesizers is to use (Σ–Δ) modulators (Razavi, 1998) to randomize the channel-selection input bits. Effectively, the phase noise and the spurs are pushed to higher frequency offsets, which are eventually sufficiently suppressed by the low-pass characteristic of the loop filter, as illustrated in Fig. 2.17.

Higher-order modulators can provide more noise suppression, but the system may become unstable. To deal with the stability issue, MASH architecture can be employed in which many stable low-order modulators are cascaded (Matsuya *et al.*, 1987).

2.5.4. *Dual-loop synthesizer*

The dual-loop synthesizer is another architecture that can achieve a high frequency resolution without the problems brought about by the low reference frequency and high division ratio of the integer-N architecture (Aytur and Khoury, 1997; Razavi,

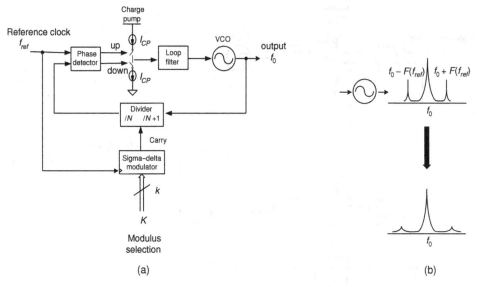

Fig. 2.17 (a) Sigma–delta fractional-N synthesizer, (b) spur with noise shaping

1997). A dual-loop synthesizer comprises two PLL systems together with a single sideband (SSB) mixer as illustrated in Fig. 2.18.

Both parallel and series configurations can be constructed to achieve the desired output frequency and frequency resolution. However, any mismatch and non-linearity of an SSB mixer will lead to spurs appearing at the mixer output. Thus, putting the SSB mixer inside the loop, as shown in Fig. 2.18(b), can suppress the spurs generated by the mixer (Kan, Leung and Luong, 2002). In both configurations, loop 1 contains a high-speed VCO with a fixed division ratio for generating the large offset frequency, while loop 2 consists of a low-speed VCO with a programmable divider for channel selection. As a consequence, loop 1 can be designed with a large loop bandwidth to suppress the close-in phase noise from its high-speed VCO, while loop 2 can be optimized with a small loop bandwidth to achieve more reduction of noise and spurs from its building blocks.

A linear model similar to that of an integer-N synthesizer can be used for a dual-loop synthesizer. For the mixer, changing its input phase is equivalent to changing its output phase. Thus, the gain of the mixer can be set to one and the linear model of the dual-loop synthesizer using series configuration is depicted in Fig. 2.19.

The basic idea behind this architecture is to add a low variable frequency from a low-frequency phase-locked loop (PLL) to a high fixed offset frequency generated from a second high-frequency PLL. Frequency change of the synthesizer therefore requires only the change of a small division ratio in the low-frequency loop. This architecture can improve the trade-off between phase noise, channel spacing,

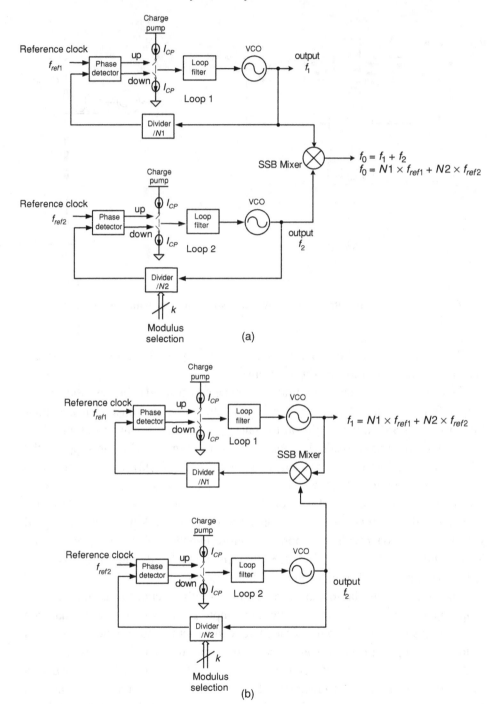

Fig. 2.18 Dual-loop synthesizer using (a) parallel configuration, (b) series configuration

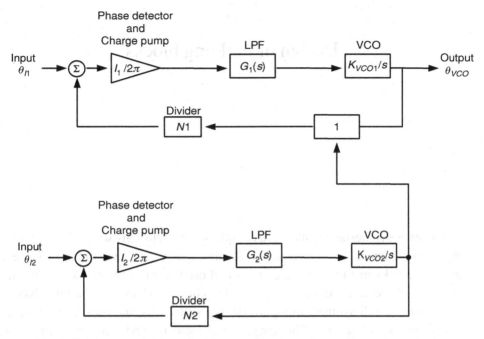

Fig. 2.19 Linear model for a dual-loop synthesizer using series configuration

reference frequency and the locking speed of the synthesizer. Although more circuits are needed, the specifications for each building block are much relaxed.

Compared with the integer-N architecture, the proposed dual-loop architecture can have much larger reference frequencies and much lower frequency-division ratios, which in turn result in a significant improvement not only to the loop bandwidth and settling time but also to the close-in phase noise. On the other hand, compared with the fractional-N architecture with sigma–delta modulation, the proposed architecture can achieve comparable performance in terms of spur and phase noise with a much simpler prescalar and without digital circuitry that would potentially introduce digital switching noise to the synthesizer.

The main drawback of the dual-loop synthesizer is that it requires two PLLs and a single sideband mixer with two reference sources. In reality, if a single reference source is preferred, the higher reference can be designed to be an integral multiple of the lower reference frequency and can be generated from the lower reference source using a third integer-N PLL.

3

Design of building blocks

A frequency synthesizer by itself is a complicated system consisting of many building blocks connected in a feedback loop. Specifically, as shown in Fig. 2.7, these building blocks include a voltage-controlled oscillator (VCO), dividers, a phase-frequency detector, a charge pump, and a loop filter. Each of these building blocks affects the overall synthesizer's performance differently and thus have different design issues and criteria. This chapter will focus on reviewing and discussing these design considerations. In addition, design issues for passive components like inductors and varactors, which are critical for high-frequency VCOs and dividers, will also be presented.

3.1. Voltage-controlled oscillators (VCOs)

One of the most critical building blocks in any phase-locked loop or frequency synthesizer is the voltage-controlled oscillator. In general, the oscillator needs to operate at the highest frequency, and its phase noise is dominant in the whole system at frequency offsets beyond the synthesizer's loop bandwidth. In current CMOS technologies, both ring oscillators and LC-tank oscillators can be fully integrated on-chip (Lee, Kim and Lee, 1997; Momtaz *et al.*, 2001), but both types have their own advantages and disadvantages. Depending on the specifications and applications, one may be more suitable than the other. The following sections will discuss and compare the design issues for the two types of VCOs.

3.1.1. Ring oscillators

Ring oscillators are typically easier to implement, and so are more desirable in standard CMOS technologies because it is not necessary to employ passive devices such as inductors and varactors. Even when these passive devices are used, the overall performance of ring VCOs is not too sensitive to the quality of the devices,

28

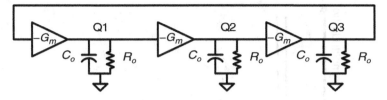

Fig. 3.1 Linear model of the three-stage ring oscillator

which is generally quite poor for standard digital CMOS processes. In addition, a large frequency tuning range capability against process variation and small chip area are other advantages of ring oscillators, which make them attractive for many designs and applications (Pottbacker and Langmann, 1994; Razavi and Sung, 1994). The main drawbacks of ring oscillators are their poor phase noise and high power consumption compared with LC oscillators.

3.1.1.1. Circuit implementation

A ring oscillator is a positive feedback system. To ensure that the system indeed oscillates, the Barkhausen criteria should be fulfilled with a total phase shift of 360° and a loop gain equal to unity (Nguyen and Meyer, 1992; Grebene, 1984, Chapter 11). Since a single-stage inverter consists of only one pole, the total phase shift of 90° is not sufficient for oscillation.

To achieve the required phase shift, several such inverters can be cascaded. For single-ended configuration, the number of cascaded stages must be odd. However, if differential structures are employed, the number of the cascaded stages can be even as long as the connections at one of the stages are reversed in polarity to obtain an extra phase shift of 180° (Buchwald and Martin, 1991). The phase shift contributed by each delay cell is dependent on the number of stages used. For example, for a four-stage ring oscillator, each stage generates a phase shift of 90°, which makes the design suitable and popular for generating in-phase and quadrature-phase outputs required for image-rejection mixers in RF transceivers and for clock recovery circuits (Razavi, 1996).

To ensure the oscillator has sufficient loop gain and total phase shift for oscillation, a linear model, depicted in Fig. 3.1, can be used to help analyze the design of a three-stage ring oscillator (Razavi, 1996).

Since each stage of the ring oscillator contains one single pole with loop gain equal to $-G_m R_o$, the transfer function can be expressed as

$$H(j\omega_{OSC}) = \left(\frac{-G_m R_o}{1 + j\omega_{OSC} R_o C_o}\right)^3. \tag{3.1}$$

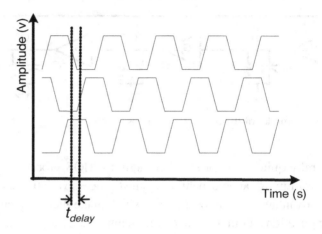

Fig. 3.2 Output waveforms of a three-stage single-ended ring oscillator

To fulfill Barkhausen's criteria, each stage contributes a 60° phase shift with the loop gain being equal to unity at the oscillation frequency:

$$\left(\frac{-G_m R_o}{1 + j\omega_{OSC} R_o C_o}\right)^3 = 1$$

$$\tan^{-1}(\omega_{OSC} R_o C_o) = 60°$$

(3.2)

and therefore,

$$G_m R_o = 2.$$

(3.3)

Once the above criteria are fulfilled, the ring oscillator can oscillate, and the output waveforms of the three stages, as shown in Fig. 3.2, are 60° out of phase with each other.

The oscillating frequency depends on the delay of each individual delay cell. Assuming that all the delay cells have the same loading, the oscillation frequency, f_o, becomes:

$$f_o \approx \frac{1}{2N t_{delay}} \approx \frac{I}{2N C_L V_p},$$

(3.4)

where N is number of stages used in the ring oscillator, t_{delay} is the time delay for each delay cell, I is the current passing through each delay cell, C_L is the total output capacitance and V_p is the peak output voltage.

It can be seen from the above discussion that the gain and phase shift of the oscillators depend on the input transconductance and the output loading. Figure 3.3 shows three types of ring oscillators with different frequency tuning mechanisms. Basically, the oscillating frequency can be tuned by (a) varying the output capacitive

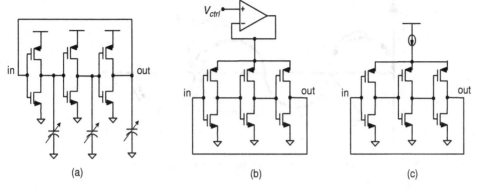

Fig. 3.3 Ring oscillator tuning by means of (a) variable capacitor, (b) supply-controlled method, and (c) current-starved method

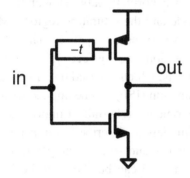

Fig. 3.4 Principle of negative skew

loading so as to change the time delay of a delay cell, (b) varying the 'on-resistance' of a linear MOSFET (Sidiropoulos *et al.*, 2000) so as to vary the RC time constant, or (c) changing the transconductance so as to change the charging and discharging time.

The oscillation frequency of ring oscillators is limited by the ratio of the device transconductance to the total parasitic capacitance at the output (G_m/C_{para}). For a given supply voltage, increasing in the device transconductance by increasing the width cannot increase the frequency significantly because the parasitic capacitance of the devices is increased as well. To improve the speed limitation of the ring oscillators, a ring oscillator using a negative skewed delay scheme, as illustrated in Fig. 3.4, has been proposed (Lee, Kim and Lee, 1997b). The conventional delay cell suffers from speed limitation because PMOS has a low mobility of *p*-type carriers. The negative skewed delay scheme can ensure that the PMOS transistors in ring oscillators are turned on before the low-to-high output transitions and turned off before the high-to-low output transitions. As such, the oscillation frequency is proved to be increased by 50% compared with the conventional approach.

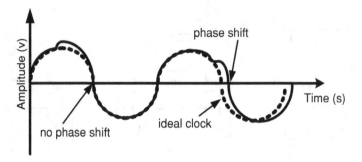

Fig. 3.5 Phase-noise effect with noise injected at different times

3.1.1.2. *Phase noise of ring oscillators*

The amplitude of oscillators is normally large enough that transistors in the oscillator work in both the triode and the saturation regions. Moreover, random noise
sources in the oscillator have periodic properties. This phenomenon is identified
as cyclostationary. Both the non-linear effect and existing cyclostationary noise
make the phase-noise analysis, which is based on a linear approach, less accurate.
The impulse-sensitivity function (ISF) can be applied to analyze the noise behavior in such non-linear phenomenon (Hajimiri, Limotyrakis and Lee, 1998). It uses
time-variant phase-noise models to describe the transient effect of perturbations
on the trajectory of oscillators and to accurately estimate the phase-noise performance of oscillators. The ISF can be shown to be large when current noise
is injected at the zero-crossing points of the oscillating signal and to be small
when the noise is injected at the peaks of the signal as illustrated in Fig. 3.5. The
phase noise becomes small or closes to zero at the peaks because the oscillator
restores the distorted signal to the initial condition before reaching the next clock
edge.

The phase noise of the three-stage ring oscillator in Fig. 3.3, based on the ISF,
can be calculated as

$$L(\Delta\omega) = \frac{3\Gamma_{rms}^2}{8\pi^2\Delta f^2} \frac{\overline{i_n^2}/\Delta f}{C_L^2 V_p^2},$$

(3.5)

where:

Γ_{rms}	is the root-mean square of the ISF;
Δf	is the offset frequency from the carrier;
$\overline{i_n^2}/\Delta f$	is the total noise power density;
C_L	is the output capacitive loading;
V_p	is the peak output amplitude.

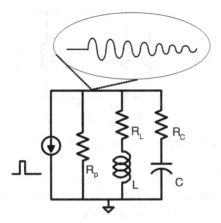

Fig. 3.6 Lossy LC passive circuit

Hajimiri's theory states that phase noise happens when oscillators have phase variation during transition. Therefore, a short transition time can help make an oscillator become less sensitive to noise disturbance and thus have good phase noise. However, more current, and thus power, would be required to reduce the transition time.

3.1.2. LC oscillators

A general criterion for designing a low-phase-noise oscillator is to minimize the number of elements required to achieve and sustain oscillation. An oscillator employing an LC tank, commonly referred to as an LC-VCO, is a good candidate to achieve oscillation at a higher frequency with lower phase noise and lower power consumption compared to the ring-type counterpart. On the other hand, the disadvantages of LC-VCOs include large chip area for the on-chip inductors and a narrow frequency tuning range.

3.1.2.1. Circuit implementation

A resonator contains an inductor and a capacitor as energy-storage elements and resistors as lossy components as shown in Fig. 3.4. Ideally, the resonator would resonate at a frequency that is mainly determined by the inductor and capacitor components, $\omega_n \approx 1/\sqrt{LC}$. However, due to the loss in the tank modeled by the resistor, the energy stored by the inductor and the capacitor is lost in every oscillating cycle, and the oscillation will eventually die out. To sustain oscillation, the loss in the LC tank should be compensated for, and this can be done by connecting the tank with a negative resistance as illustrated in Fig. 3.6.

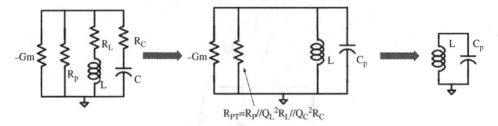

Fig. 3.7 Lossy LC passive circuit with negative conductance

In advanced sub-micron CMOS technologies, the quality factor of the capacitor is usually high (more than 30) so the capacitance value in a parallel configuration is similar to that in a series configuration, which is given by Equation (3.6):

$$C_p = C\frac{Q_c^2}{Q_c^2 + 1} \approx C \tag{3.6}$$

The loss in the tank is dominantly contributed by the loss of the on-chip inductor as illustrated in Fig. 3.7. The admittance of the LC tank can be described as

$$Y(j\omega) = \frac{1}{j\omega L} + j\omega C_p + \frac{1}{R_{pT}} - G_m \tag{3.7}$$

where R_{pT} is the equivalent parallel output impedance at each of the output nodes, which is typically dominated by the parallel output impedance of the inductor itself.

At the resonance frequency, the imaginary and real part are zero, thus the frequency of the oscillator can be expressed as

$$imag[Y(j\omega_0)] = 0 \tag{3.8}$$

$$\omega_0 \approx \frac{1}{\sqrt{LC_p}}.$$

To be exact, the lossy LC tank in practice oscillates at a frequency slightly lower than that of the tank formed by lossless inductor and capacitor.

To guarantee that the oscillation occurs, the negative conductance $-G_m$ should be designed so that

$$real\{Y(j\omega_0)\} \leq 0 \tag{3.9}$$

$$G_m \geq \frac{1}{R_{pT}} = \frac{1}{R_L(1 + Q_L^2)},$$

where Q_L is the quality factor of the inductor defined as

$$Q_L = \frac{\omega L}{R_L}. \tag{3.10}$$

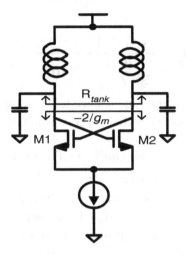

Fig. 3.8 VCO with cross-coupled transistors

To provide the required negative transconductance for a differential lossy resonator, a simple circuit can be used, as shown in Fig. 3.8. It consists of a cross-coupled differential pair formed by M1 and M2 for which the single-ended impedance $-G_m$ is derived and shown in Equation (3.11):

$$-G_m = -g_m, \tag{3.11}$$

where g_m is the transconductance of the transistor.

By considering the loss in the LC tank, the required transconductance should be

$$g_m \geq \frac{1}{R_{pT}} = \frac{1}{R_L(1 + Q_L^2)} \approx \frac{1}{(\omega_o L)Q_L} = \frac{\omega_o C}{Q_L}. \tag{3.12}$$

However, to guarantee that the VCO oscillates, the transconductance is usually designed to be about two times larger than the required value.

Once the required transconductance is known, the power dissipation of the VCO can be derived and is given by

$$\begin{aligned} \text{Power} &= (I_{M1} + I_{M2})V_{DD} \\ &= 2I_{M1}V_{DD} \end{aligned} \tag{3.13}$$

and

$$I_{M1} = \mu C_{ox}\frac{W}{2L}(V_{GS} - V_T)^2 = \frac{g_m^2}{2\mu C_{ox}\dfrac{W}{L}}, \tag{3.14}$$

where C_{ox} is the oxide capacitance per unit area, and V_T is the threshold voltage of the devices.

Therefore,

$$\text{Power} = \frac{V_{DD}}{\mu C_{ox}\dfrac{W}{L}}\frac{1}{R_{pT}^2} = \frac{V_{DD}}{\mu C_{ox}\dfrac{W}{L}}\frac{C}{LQ_{L^2}} \tag{3.15}$$

This clearly illustrates that larger inductance-to-capacitance ratio L/C value and higher quality factor can minimize the power consumption of the VCO.

3.1.2.2. Phase noise of LC oscillators

An LC oscillator is able to generate a signal with high spectral purity due to the band-pass characteristic of the resonant tank. The quality factor of the LC tank determines the ratio of the oscillator's signal to its noise spectral density. Equation (3.16) shows the relation between phase noise and other parameters of an LC oscillator (Craninckx, 1995).

$$L(\Delta\omega) = 10 \log\left[KTR_{eff}(1 + F)\left(\frac{\omega_o}{\Delta\omega}\right)^2\frac{1}{V_{rms}^2}\right], \tag{3.16}$$

where:

K	is Boltzmann's constant with a value of 1.38×10^{-23} J/K;
T	is the absolute temperature;
F	is an excess noise factor;
R_{eff}	is the LC tank's effective resistance;
ω_o	is the oscillation frequency;
$\Delta\omega$	is the offset frequency from ω_o;
V_{rms}	is the rms differential output amplitude.

Equation (3.16) indicates that the series resistance of the on-chip inductors in the tanks should be as small as possible. At the same time, the amplitude of the VCO should be as large as possible. However, the oscillating amplitude is bounded by the supply voltage and cannot be increased infinitely. There are two regimes to characterize the working phenomenon of the LC oscillator (Hegazi, Sjoland and Abidi, 2001). As long as the output amplitude is controlled by the bias current source of the cross-coupled differential pair, the VCO works in the current-limited regime. If the bias current is increased such that the amplitude becomes limited by the voltage supply, the VCO is pushed into its voltage-limited regime. In the voltage-limited regime, the negative peaks of the signal are strong enough to force the bias current source into a triode region and one of the g_m devices to be off. Non-linear behavior of the transistors causes the baseband noise to be converted to the oscillation frequency through the commutating switches of the negative transconductance cell. Therefore, the phase noise becomes worse, even though the current and the power consumption are increased.

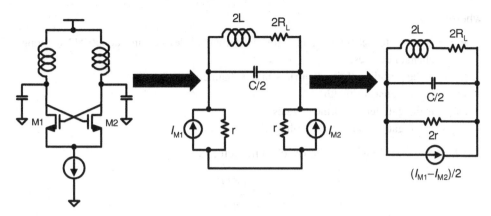

Fig. 3.9 Transformation of the differential LC oscillator to its equivalent noise model

In the current-limited regime, the sideband noise of the oscillators is mainly limited by the loss in the tank. For higher loss, a larger transconductance is needed to be able to compensate and sustain the loop gain for a stable oscillation. Therefore, it is always beneficial to use a tank with the quality factor (Q) as high as possible. Nevertheless, the Q-value of on-chip planar inductors in the current CMOS processes is typically limited to around 5 to 10. One solution to increasing the Q of the tank to improve the phase noise for a given power or to minimize the power for a given phase noise is to use bond wires in place of on-chip inductors (Craninckx and Steyaert, 1995b). Since the quality factor of bond wires can be as high as 40, the VCO performance could be significantly improved. However, such a solution is not attractive due to the problems with controllability, repeatability, and reliability of the bond wires.

3.1.2.3. Phase-noise analysis

Noise analysis by the linear method shown in Equation (3.16) can give a first-order approximation of the phase-noise performance of LC oscillators. A more accurate phase-noise analysis can be based on a linear time-varying model proposed by Hajimiri as discussed in Section 3.1.1.2. The transformation of the differential LC oscillator to its equivalent noise circuit model is shown in Fig. 3.9.

Thus, the total noise power density of the cross-coupled transistors $\overline{i_{cc}^2}$ is

$$\overline{i_{cc}^2} = 2\frac{\overline{i_1^2}}{4} = 2KT\gamma\mu_n C_{ox}\frac{W_n}{L_n}(V_{GS} - V_T), \tag{3.17}$$

where:

γ is equal to between 2 and 3 for short-channel devices and about 2/3 for long
        channel devices;

μ_n is the mobility of the *n*-carriers in the channel;

W_n is the transistor width;

L_n is the gate length of the devices;

V_{GS} is the gate-to-source DC voltage.

The noise power density of the on-chip inductor is

$$\frac{\overline{i_L^2}}{\Delta f} = 2\frac{4KTR_L}{\omega_o L}, \tag{3.18}$$

where:

R_L is the series resistance of a inductor including metal loss, skin effect and substrate
        loss;

L is the inductance value;

ω_o is the oscillating frequency.

Thus, the phase noise can be calculated from

$$L(\Delta\omega) = \frac{2\Gamma_{rms}^2}{8\pi^2\Delta f^2 C_L^2 V_p^2}\left[KT\gamma\mu_n C_{ox}\frac{W_n}{L_n}(V_{GS} - V_T) + \frac{4KTR_L}{\omega_o L}\right]. \tag{3.19}$$

3.1.2.4. *Quadrature oscillators*

Typical differential oscillators have two outputs that are 180° out of phase with each other. However, in some applications, four or even more output phases may be needed for data processing. As an example, four differential in-phase and quadrature-phase outputs are commonly required in RF transceivers for image-rejection mixers. Ring oscillators are capable of generating multiple clock phases at the outputs of their delay cells. However, it will result in a reduction of the oscillating frequency as the delay is directly proportional to the number of stages in the loop. As discussed earlier, ring oscillators inherently have low operation frequency and poor phase noise. A polyphase filter consisting of resistors and capacitors, RC–CR, can be employed at the outputs of the VCO to generate multiple-phase outputs. However, the phase errors between the outputs can be small for only a small frequency range. Using a high-order polyphase filter can achieve a larger bandwidth with better matching (Tadjpour *et al.*, 2001). To minimize the mismatches and to reduce the phase errors at the outputs, components need to be large, which would, in turn, not only occupy larger chip area, and consume larger power consumption, but also contribute more noise. To overcome this ambiguity, quadrature LC oscillators (Rofougaran *et al.*, 1996) or multiphase LC oscillators (Kim and Kim, 2000) can

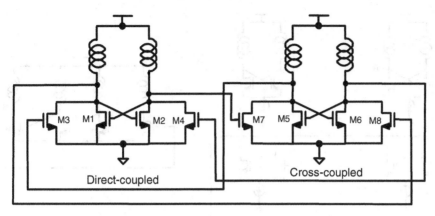

Fig. 3.10 Quadrature LC oscillator

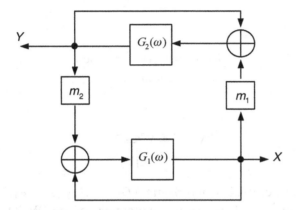

Fig. 3.11 A linear model of a quadrature oscillator

be employed. Basically, these oscillators use two or multiple LC resonant tanks as individual delay cells in a ring configuration as in a ring oscillator.

A schematic of the quadrature LC oscillator (QVCO) is shown in Fig. 3.10. Two identical LC differential VCOs are connected in direct-coupled and cross-coupled configuration.

In order to illustrate that this configuration has a phase difference of 90°, Fig. 3.11 shows a linear model for analyzing the behavior of the QVCO (Liu, 1999). $G(\omega)$ represents the open-loop gain of each VCO, and the parameter m corresponds to the coupling coefficient between the tanks. In the steady state, the two oscillators in the loop have the same oscillation frequency. Therefore, the output phase of each VCO can be written as

$$(X + m_2 Y)G_1(j\omega) = X \tag{3.20}$$

$$(Y + m_1 X)G_2(j\omega) = Y. \tag{3.21}$$

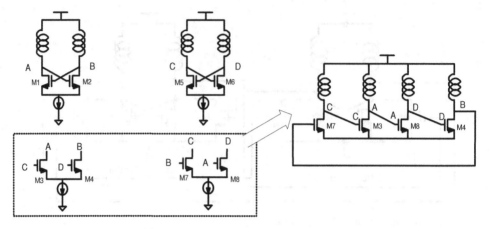

Fig. 3.12 Redrawn schematic of the QVCO

Assume that

$$G_2(j\omega) = G_1(j\omega) = G(j\omega)$$
$$m_1 = -m_2 = m \tag{3.22}$$

then,

$$X^2 + Y^2 = 0$$
$$X = \pm jY, \tag{3.23}$$

where the coupling coefficient m is the ratio G_{M3-4}/G_{M1-2}, or G_{M7-8}/G_{M5-6}.

The above equations show that the two outputs of the QVCO have the same amplitude but a 90° phase difference. In other words, the oscillator indeed achieves both in-phase and quadrature-phase output signals.

The analysis shows that the quadrature oscillator has two phase patterns. However, in the practical situation, only one pattern appears. To facilitate the analysis of the phase relation, Fig. 3.12 depicts a redrawn schematic of the QVCO.

The two possible patterns are illustrated in Fig. 3.13 in the time domain together with their phasor diagrams.

In fact, Node A in Fig. 3.12 is determined by the NMOS transistors M1 and M3, which depend on voltage V_B and V_C, respectively. Therefore, the resultant current I_A flowing into the LC tank at node A is given by:

$$I_A = -g_{M1} \cdot \bar{V}_B - g_{M3} \cdot \bar{V}_C, \tag{3.24}$$

where g_{M1} and g_{M3} are the large-signal transconductances of the devices.

Assuming that g_{M1} and g_{M3} are equal to g_M, and the four output phases of the QVCO are the same in the steady-state oscillation and that $|V_B| = |V_C| = V$,

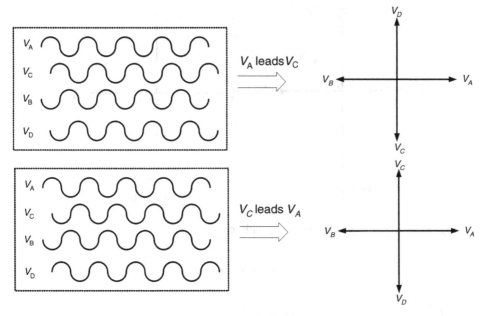

Fig. 3.13 The two possible time waveforms of the QVCO and their phasor diagrams

Equation (3.24) can be expressed as:

$$I_A = g_M \cdot V[\cos(\omega t) + j\sin(\omega t)], \tag{3.25}$$

when V_A leads V_C or

$$I_A = g_M \cdot V[\cos(\omega t) - j\sin(\omega t)], \tag{3.26}$$

when V_C leads V_A.

The phasor diagrams of I_A with respect to V_A are shown in Fig. 3.14. On the other hand, for a parallel LC resonator, the impedance reaches its maximum peak when its phase shift is around zero. The frequency becomes lower and higher at the negative and positive phase shifts, respectively, as illustrated in Fig. 3.15. The figure also indicates that the impedance at the negative phase shift is higher than that at the positive phase. Because a higher impedance results in a higher loop gain, only the case of V_A leading V_C is actually possible in practice.

3.1.2.5. Frequency tuning

The resonance frequency of an LC oscillator is mainly determined by the inductance and capacitance in the resonant tank. Varactors are widely used for varying capacitance. These can be implemented by using *pn* junctions or MOS varactors as will be described later in Section 3.8. Since varactors can only be used for fine and

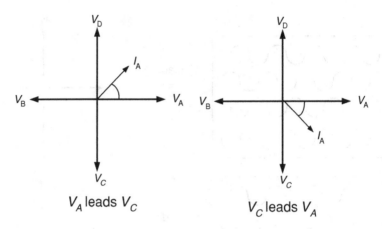

Fig. 3.14 The phasor diagrams of I_A with respect to V_A

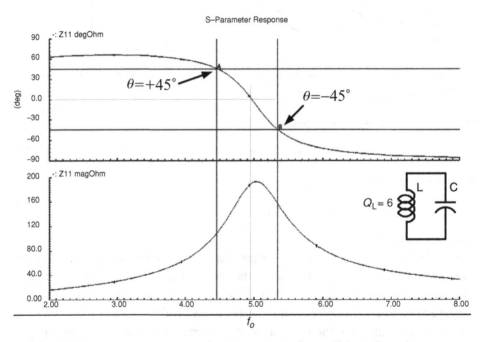

Fig. 3.15 Impedance and phase of an LC resonator in parallel configuration

limited tuning, a step and coarse tuning scheme may be more desirable to cover a wide frequency tuning range to compensate for process variation. Such a coarse tuning can be achieved by using switched-capacitor arrays (SCAs), which are composed of fixed value capacitors in series with MOS switches. Different combinations of the switches can be turned on and off to switch in different capacitance to achieve different coarse oscillation frequencies.

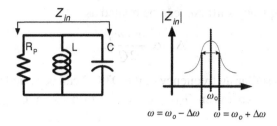

Fig. 3.16 Parallel RLC tank

The disadvantage of using variable capacitors is their large parasitic capacitance, which inevitably limits the frequency tuning range. As an alternative approach to tuning capacitance, inductive tuning is proposed by connecting MOS switches in series with inductors (Andreani, 2001; Herzel, Erzgraber and Ilkov, 2000). This technique can achieve a large frequency tuning range but at the expense of the phase noise due to extra loss from the turn-on resistance of the switches, which is quite critical and renders it highly undesirable.

Based on the previous analysis, another tuning mechanism can be adopted by varying the coupling coefficient m of the quadrature oscillators. Based on Equations (3.20), (3.21) and (3.22), the following equation can be obtained:

$$G(\omega) = \frac{1}{(1 \pm jm)}. \tag{3.27}$$

Therefore,

$$\phi(G(\omega)) = \pm \tan^{-1} m \tag{3.28}$$

For a parallel LC resonator, as shown in Fig. 3.16, the impedance Z_{in} can be derived as

$$Z_{in} \approx \frac{1}{\dfrac{1}{R_p} + \dfrac{1}{j(\omega_o \pm \Delta\omega)L} + j(\omega_o \pm \Delta\omega)C}$$

$$\approx \frac{1}{\dfrac{1}{R_p} + \dfrac{\pm j\Delta\omega}{\omega_o^2 L} + j(\omega_o \pm \Delta\omega)C} \tag{3.29}$$

$$\approx \frac{R_p}{1 \pm 2j\dfrac{Q\Delta\omega}{\omega_o}}.$$

Therefore, the frequency shift $\Delta\omega$ can be related as

$$\Delta\omega \propto \frac{m \cdot \omega_o}{2Q}. \tag{3.30}$$

It is clear that the oscillating frequency of the QVCO can be varied by changing the transconductance ratio between G_{M3-4} and G_{M1-2} or between G_{M7-8} and G_{M5-6}.

3.1.2.6. Amplitude modulation to phase modulation noise

As discussed previously, it is quite common to use varactors to achieve fine tuning of the oscillation frequency of an LC oscillator. The envelope and the duty cycle of the oscillating waveform determine the average capacitance of the varactors. Since the amplitude modulation (AM) noise of the oscillating signal inevitably modulates the average capacitance of the varactors, the oscillating frequency is actually shifted, and consequently, sidebands are generated (Hegazi, Sjoland and Abidi, 2001). This effect becomes more severe when the oscillator needs to have a large VCO gain because varactors with larger capacitance tuning ratios are required.

3.2. Dividers

Nowadays, the operating frequency of VCOs designed in CMOS technology can be as high as 50 GHz (Wang, 2001). To be able to track high-frequency output signals from such VCOs, to lock PLLs or synthesizers, it is important to include in the feedback loop CMOS frequency dividers that can operate at the same operation frequencies. On the other hand, after the first few dividers, the signal frequency becomes low enough that it is more power efficient to use other dividers that can operate at a much lower frequency but with much lower power consumption. As such, both divider topologies are generally needed, one of which is for high frequencies but high power consumption, whereas the other is for low frequencies and low power consumption. The following sections will describe different divider implementations for the two categories.

3.2.1. Source-coupled logic divider with resistive load

The differential source-coupled logic (SCL) frequency divider is generally recognized as the fastest divider topology and it can be realized by cascading two D-latch stages as shown in Fig. 3.17. Two D-latch stages are cascaded with the output of the second stage cross-coupled to the input of the first stage to perform a divide-by-2 function. Each of the two D-latches consists of a cross-coupled pair (M3, M4) connected in a positive feedback configuration to provide negative transconductance to maximize the operation frequency. Each D-latch is driven by a single clock with two

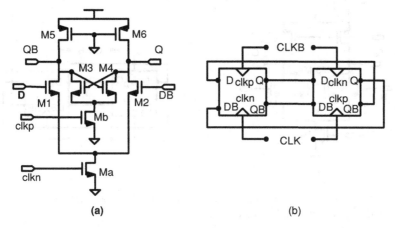

(a) **(b)**

Fig. 3.17 SCL frequency divider (a) D-latch, (b) divide-by-2 using D-latches

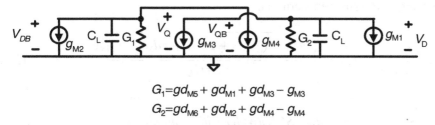

$$G_1 = gd_{M5} + gd_{M1} + gd_{M3} - g_{M3}$$
$$G_2 = gd_{M6} + gd_{M2} + gd_{M4} - g_{M4}$$

Fig. 3.18 Small-signal model of D-latch

complementary clock phases, one of which is used to control the flipping circuits formed by M1 and M2, and the other to control the latching circuits formed by M3 and M4. The pull-up network formed by PMOS devices M5 and M6 is connected from V_{DD} to the outputs as the load. Once the clock signal is high, M1 and M2 start switching on and off based on the differential input signals D and DB. One of the output nodes, Q or QB, is discharged through the latching circuits. The opposite output node is charged up as the signal path through PMOS, providing a relative low impedance. When the clock signal is low, M1 and M2 are off, while M3 and M4 latch the outputs and keep the state of the outputs until the clock signal is high again.

To achieve maximum speed with minimum power consumption, the devices used in the latches should be carefully designed. Based on the small-signal model shown in Fig. 3.18, the corresponding transfer function of the D-latch can be obtained:

$$A(s) = \frac{V_Q}{V_D} = \frac{g_{M1}}{gd_{M5} + gd_{M1} + gd_{M3} + sC_L - g_{M3}}; \qquad (3.31)$$

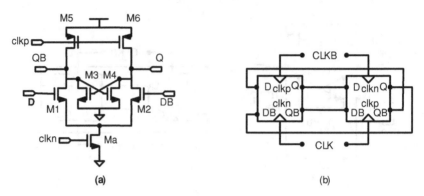

Fig. 3.19 SCL divider with dynamic loading (a) D-latch proposed by Wang (2000), (b) divide-by-2 using the D-latches

and the output frequency f_{out} can be approximated by setting $A(s)$ equal to unity according to Barkhausen's criteria:

$$f_{out} = \frac{\sqrt{g_{M1}^2 - (gd_{M5} + gd_{M1} + gd_{M3} - g_{M3})^2}}{2\pi C_L}. \tag{3.32}$$

3.2.2. SCL divider with dynamic load

As mentioned before, SCL dividers operate in two different modes: a latching mode and a flipping mode. In the flipping mode, the RC time constant should be as small as possible to shorten the transitions and thus to maximize the speed. On the other hand, in the latching mode, the RC time constant should be as large as possible to achieve maximum gain for good latching. As a result, the operating frequency can be increased by dynamically changing the output loading with respect to the operating modes. This approach is commonly referred to as a dynamic-loading frequency divider as shown in Figs. 3.19 (Wang, 2000; Wong, Cheung and Luong, 2002). In such a divider design, the gates of the PMOS loading devices are connected to a clock signal (*clkp*) as opposed to a constant bias as in the conventional divider. The complementary clock signals *clkp* and *clkn* are used to drive the gate of PMOS loading devices and the gate of the bias current device Ma, respectively. When *clkp* is low and *clkn* is high, PMOS devices M5 and M6 are operated in their linear region and have a small impedance, which results in a small RC time constant for sensing the input signal rapidly. When *clkp* is high and *clkn* is low, PMOS devices M5 and M6 have a small gate-to-source voltage and the impedance becomes larger, which enables the negative G_m-cell formed by M3 and M4 to have a large gain to flip the outputs.

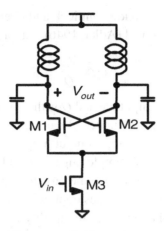

Fig. 3.20 Injection-locked frequency divider

3.2.3. Injection-locked frequency divider

Equation (3.31) indicates that the open loop gain can be made larger by strengthening the negative transconductance of the cross-coupled pair, resulting in speeding up the divider. Moreover, the RC time constant contributed by $1/g_{M5-6}$ and C_L can be reduced to increase the operating frequency. Since the imaginary part of the output impedance cannot be eliminated by the negative tranconductances alone, it requires inductance at the load to resonate out the capacitive load for achieving a higher operating frequency. As a result, the transfer function of a divider using inductor load in the tank can be expressed as

$$A(s) = \frac{V_Q}{V_D} = \frac{g_{M1}}{gd_{M5} + gd_{M1} + gd_{M3} + sC_L + \frac{1}{sL_L} - g_{M3}}. \qquad (3.33)$$

With an inductive load, a single-stage D-latch is sufficient to provide a total phase shift of 360°. In fact, the circuit is the same as for an LC oscillator but can be used as a divider by injecting an incident signal to the common source node of M1 and M2. This kind of divider is called an injection-locked frequency divider (Rategh and Lee, 1999), and a schematic is shown in Fig. 3.20. With an input signal at a frequency ω_1 injected to the common node, the output will be sinusoidal at a frequency $\omega_1/2$. The bias of the current source is from the common mode of the input. Without injecting a signal into the current source of the oscillator, the DC bias current may still be able to sustain the oscillation in which the divider acts as an LC oscillator with the oscillation frequency depending on the inductors and the total capacitors at the output.

However, using inductors as the load, injection-locked dividers occupy a much larger chip area and at the same time have a limited locking range, which is the frequency range in which the divider can lock and operate properly. The locking range $\Delta\omega$ based on Adler's model (Adler, 1946) is approximately related by

$$\Delta\omega \propto \frac{V_{in}}{Q}, \tag{3.34}$$

where V_{in} is the input signal amplitude and Q is the quality factor in the tank.

Thus, using inductors with a smaller Q can achieve a larger locking range, but using a larger Q can have a larger output amplitude. As a consequence, the divider design can be optimized by employing low-Q inductors but at the same time increasing the bias current to maximize the output amplitude. This unfortunately would result in an increase in power consumption.

It is interesting to note that although the quality factor in the LC tank can be designed to be small, the phase noise is not degraded when it is locked to the input signal. As a matter of fact, the injection-locked divider is a first-order PLL. The output phase of the divider will follow closely the input phase. Therefore, the phase noise of the divider is mainly determined by the phase noise of the input signal rather than the noise of the divider itself.

3.2.4. *True single-phase clock divider*

A true single-phase clock (TSPC) divider based on TSPC D-type flip-flops is a digital frequency divider, which is generally simpler than its analog counterparts like SCL and injection-locked dividers. The TSPC D-type flip-flop was first proposed by Yuan and Svensson (1989) and, as shown in Fig. 3.21(a), requires only one single clock phase and contains only nine transistors. Owing to the small number of transistors and the small delay from 'D' to 'out', the operation frequency can be high. Therefore, the TSPC divider is a more suitable architecture for a divider than a traditional static divider when working at high frequency. However, the TSPC divider requires the input clock to have a nearly rail-to-rail voltage swing in order to achieve high frequency operation. Applications with the same or higher operation frequency can be achieved by the SCL architecture, which does not require a rail-to-rail input signal swing but has more power consumption. On the other hand, if the frequency of the input clock is too slow, the rates of charging and discharging are also too slow, and leakage will change the voltages of the internal nodes and result in glitches, which in turn, will cause a wrong division ratio.

The operation of the TSPC D-type flip-flop is as follows and consists of two working modes: the evaluation mode and the hold mode. When CLK is high, the D-type flip-flop works in the evaluation mode. If node $n1$ is high, the transistors mn2 and mn3 are turned on. Node $n2$ will be pulled low, and OUTB becomes high. If node $n1$ is low, the transistor mn2 is turned off, and node $n2$, which was previously

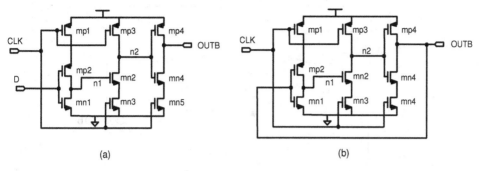

Fig. 3.21 (a) TSPC D-type flip-flop, (b) divide-by-2 implementation

pre-charged, remains high. Thus, the state of OUTB becomes low. Therefore, node $n1$ is transparent to the output OUTB in the evaluation mode. When CLK is low, the D-type flip-flop works in the hold mode, namely the pre-charged mode. Node $n2$ is pre-charged to high through mp3, and transistors mn3 and mn5 are off, the value at OUTB is held.

Figure 3.21(b) illustrates the implementation of a TSPC divide-by-2 circuit, which has the clock input as the output of the previous stage and the output being fed back to the D input. As the value at node $n1$ is the inverted value of the input D, when the data at node $n1$ is transmitted to OUTB in the evaluation mode, the data at OUTB becomes the inverted value of the input D. When the output node OUTB is fed back to D, OUTB will toggle its own state after two clock cycles. Hence, it can perform the divide-by-2 function.

3.2.5. Divider using static logic

Since the TSPC D-type flip-flop utilizes dynamic logic, the voltage at each node of the D-type flip-flop is stored by the parasitic capacitors of the transistors. Leakage of transistors can change the states of the operating point if the period of the input clock is too long, and the operation of divide-by-2 may eventually fail. The TSPC divider is good for high-frequency operation, but it cannot work at very low frequency. At an operating frequency where the speed is not critical, a static logic design, as shown in Fig. 3.22, can be employed for low power and glitch-free consideration.

3.3. Prescaler

A prescaler is a combination of several simple dividers and counters to achieve a sophisticated divider with more complicated division ratios. In practice, different divider and prescaler topologies can be combined in order to achieve both the

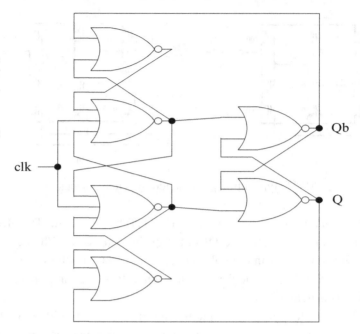

Fig. 3.22 Frequency divider using static logic

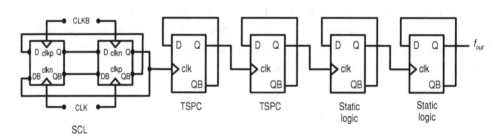

Fig. 3.23 Example of divide-by-32 using a cascade of several dividers

required speed and the total division ratio while minimizing the power consumption. Moreover, depending on the specification and applications, prescalers may be either non-programmable or programmable, the latter of which is required to obtain different division ratios for channel selection.

3.3.1. Non-programmable prescaler

Non-programmable dividers have fixed division ratios and can simply be constructed by cascading several dividers in series as illustrated in Fig. 3.23. After the first-stage divider, the successive dividers are configured as a ripple counter to pass signals through one stage after another. As a result, the operation frequency is

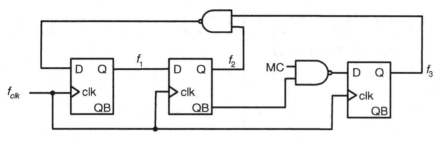

Fig. 3.24 Block diagram of a divide-by-4/5 circuit

the highest for the first divider but becomes lower and lower from stage to stage. The first few dividers can be realized employing either SCL or injection-locked topologies as discussed in Section 3.2 to achieve the desired high operation frequency. The last few dividers operating at lower frequency can be implemented using TSPC or static-logic dividers to minimize their power consumption and complexity.

3.3.2. Dual-modulus prescaler

In order to change the VCO frequency without varying the reference frequency in a synthesizer, the division ratio of the divider in the feedback loop should be changed, which can be accomplished by a programmable prescaler. Designing a programmable prescaler is more difficult than designing a fixed-division-ratio prescaler, because the division ratio should be toggled fast enough for high-frequency operation.

As discussed in Chapter 2, all PLL-based synthesizer architectures require a dual-modulus prescaler, which is capable of being selected by a modulus control signal to divide by N or $N + 1$. A conventional design of a synchronous divide-by-4/5 is illustrated in Fig. 3.24 (Rogenmoser, Huang and Piazza, 1994).

The dual-modulus prescaler makes use of several combinational circuits to obtain the modulus control signal MC for selecting the desired division ratios. As shown in the timing diagram (Fig. 3.25), when MC is zero, the value of f_3 is always high, NAND2 and D3 become inactive, and NAND1 functions as a NOT gate. Flip-flops D1 and D2 form a Johnson counter, and the whole prescaler performs a divide-by-4 function. When MC is high, NAND2 works as a NOT gate. Thus, when both f_2 and f_3 are equal to 1, the output of NAND1 is 0. The value of f_1 changes from high to low after three consecutive high values of f_{clk}. As a result, the prescaler performs a divide-by-5 function.

Design of the combinational circuits at high operation frequency with minimum power consumption becomes critical for the prescaler design. Another prescaler

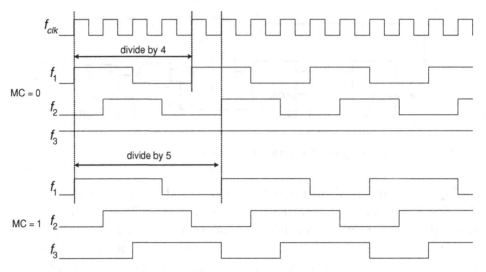

Fig. 3.25 Output waveforms of synchronous divide-by-4/5

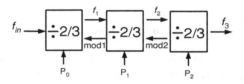

Fig. 3.26 Three-bit multi-modulus prescaler using divide-by-2/3

topology using back-carrier-propagation to implement a dual-modulus divider-by-8/9 has been reported by Larsson (1996). The input frequency was improved from 1.7 GHz to 1.9 GHz using a 0.8 μm CMOS technology at 5 V. Yet, another prescaler using phase-switching architecture was proposed with a maximum input frequency up to 2.65 GHz at 5 V in a 0.7 μm CMOS process (Craninckx, 1996).

3.3.3. Multi-modulus Prescaler

Some synthesizer implementation may require a multi-modulus prescaler that can provide more than two division ratios. As an extension of dual-modulus versions, such multi-modulus prescalers can be designed by cascading several dual-modulus prescalers. A three-bit multi-modulus prescaler is illustrated in Fig. 3.26 (Vaucher *et al.*, 2000). The multi-modulus prescaler is controlled by three selection digital bits, each of which is used to control one of the three dividers-by-2/3 to divide by 2 or 3 when it is low or high, respectively. As given by Equation (3.35), the total

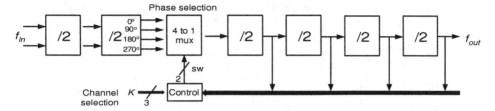

Fig. 3.27 Multi-modulus prescaler using phase-switching techniques (Craninckx and Steyaert, 1998)

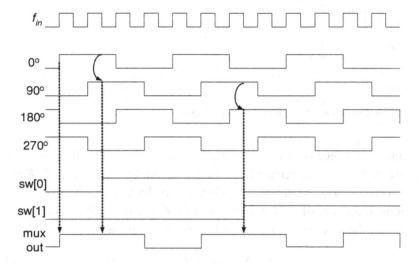

Fig. 3.28 Timing diagram of the prescaler using phase-selection circuit

division ratio can be selected to be any number from 8 to 15.

$$N = 2^3 + P_2 \cdot 4 + P_1 \cdot 2 + P_0 \tag{3.35}$$

Designing a finite state machine to control the required combination logic at a high frequency with low power consumption is challenging. Such a finite state machine can be eliminated by employing phase-switching techniques (Craninckx and Steyaert, 1998). As shown in Fig. 3.27, the first two stages are divide-by-2 circuits, and phase switching is performed at the frequency of $f_{in}/4$ to relax the requirement and power dissipation of the prescaler circuits.

One criterion of this prescaler is that the second-stage divider should be able to generate four output phases that are 90° out of phase with each other. The phase-selection circuit, upon insertion of the control signal bits, chooses one of the four input phases as its output. When the next control signal bits are applied, the phase-selection circuit selects the next phase as its output as shown in Fig. 3.28. Therefore,

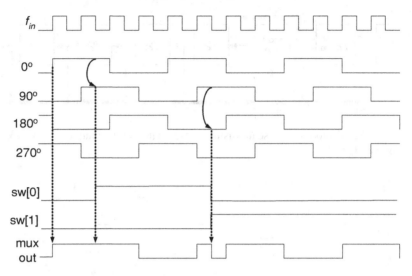

Fig. 3.29 Phase switching with glitch problem

the selection sequence is from $0°$ to $90°$, $180°$, and $270°$, and returns to $0°$ again. The phase switching is determined by both the channel-selection signals K and the outputs of the subsequent divide-by-2 circuits. As the phase switching is in steps of $90°$ at the frequency of $f_{in}/4$, the output of the 4-to-1 multiplexer corresponds to one period of the input signal f_{in}. The division ratio is given by

$$N = 64 + K, \tag{3.36}$$

where K is ranged from 0 to 7.

The disadvantage of using the phase-switching approach is the possible presence of spikes during the switching process, as shown in Fig. 3.29. For example, if the transition occurs when the signal of $90°$ is high and the signal of $180°$ is low, the output of the 4-to-1 multiplexer would appear with glitches. These glitches can trigger the successive divider and lead to wrong division ratios. To remedy this problem, retimed multiplexer control signals, which use digital logic circuits to control when the phase-switching process actually takes place, are introduced by Krishnapura and Knight (2000). The scheme ensures that both phases are either 1 or 0. In other words, the system avoids the transition from 1 to 0 or 0 to 1. However, the power consumption using this technique increases substantially. Another technique that can prevent the glitch problem is backward phase selection (Shu *et al.*, 2003). The sequence of phase switching is from $270°$, $180°$, $90°$, $0°$, instead of from $0°$, $90°$, $180°$, $270°$ (Fig. 3.30). Even though the transition is from 1 to 0 or 0 to 1, no glitch would appear during phase transitions.

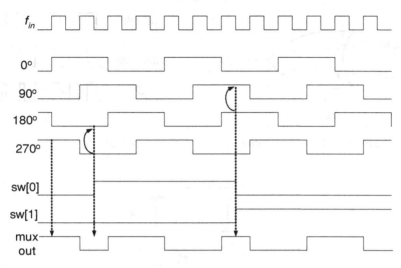

Fig. 3.30 Phase switching using backward phase selection to prevent glitch problem

3.4. Phase–frequency detectors (PFDs)

A phase detector (PD) detects the phase difference at its inputs and generates corresponding 'up' and 'down' outputs to control charge pumps. A PD is normally able to work when two input signals have a very small frequency difference. Once the frequency difference gets large enough, another frequency-locked loop or a phase-frequency detector (PFD) is needed to perform phase and frequency comparisons. In general, a PFD can offer a larger acquisition range than a simple PD.

3.4.1. Phase detector design

There are two main types of phase detector (PD) implementation, one is for analog PD and the other is for digital PD. Analog PDs in general can operate at higher frequencies than digital PDs and can be realized simply using Gilbert cell architecture (Gray and Meyer, 1992), in which the output depends on the phase error of the input signals. However, the gain of these PDs depends on the input signal amplitude, which in turn affects the loop gain of the whole PLL. In addition, the power consumption is relatively high compared to digital PDs. Digital PDs can be implemented with exclusive-OR (XOR) logic gates as illustrated in Fig. 3.31. This architecture is simpler and the gain is independent of the input amplitude. On the other hand, since they are digital circuits, the input signals are required to be rail-to-rail, and the operating speed is therefore slower.

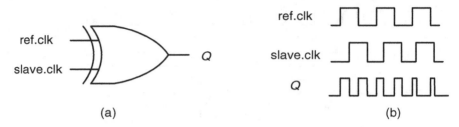

Fig. 3.31 (a) XOR phase detector, (b) timing diagram of (a)

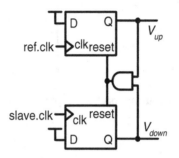

Fig. 3.32 Block diagram of a PFD

3.4.2. Phase–frequency detector design

A phase–frequency detector (PFD) can perform both phase and frequency comparisons. A generic block diagram of a PFD is shown in Fig. 3.32, which consists of two D-type flip-flops where the D inputs are always high.

The PFD is basically a finite-state machine. As illustrated in Fig. 3.33 and Fig. 3.34, the pulse widths of the V_{up} and V_{down} signals are changed proportionally according to the phase difference between the reference clock and the slave clock.

As illustrated in Fig. 3.35, a PFD can be constructed with two TSPC D-type flip-flops and a NOR gate in the feedback path for the reset.

The D-type flip-flops are designed to be triggered at rising edges with the D inputs being constantly high. The logic outputs, V_{upB} and V_{downB}, are determined by the reset and the clock signals, *ref.clk* and *slave.clk*. The outputs go high on a rising clock edge as long as the reset is low. The outputs remain high until the reset goes high. When the two clocks signals are simultaneously high, the reset changes its status from low to high, thereby resetting both the output signals, V_{upB} and V_{downB}.

When the rising edge of the reference signal is faster than the slave signal, the pulse width of the 'up' output is wider than the 'down' output. The net charge in the loop filter is increased and, as a result, the frequency of the VCO is increased. When

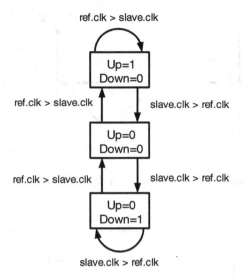

Fig. 3.33 Finite state machine of a PFD

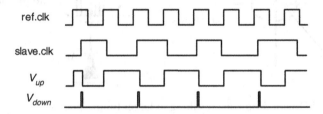

Fig. 3.34 Timing diagram of a PFD

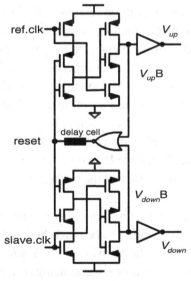

Fig. 3.35 PFD using TSPC D-type flip-flop

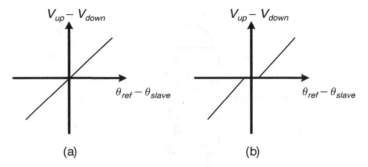

Fig. 3.36 PFD transfer function (a) without dead zone, (b) with dead zone

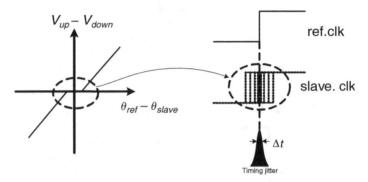

Fig. 3.37 PFD with dead zone resulting in larger timing jitter

the slave signal becomes faster than the reference signal, the 'down' output has a larger pulse width and discharges the loop filter to decrease the VCO frequency.

A delay cell is inserted after the NOR gate in the feedback path to ensure that, when the PLL is in steady state, the reset function has sufficient time to maintain 'up' and 'down' pulses. This is essential to prevent a dead-zone problem in which the phases of the two input signals are very close to each other but synchronization cannot be reached, and the PFD takes no action for phase adjustment. Figure 3.36 illustrates that the output of a PFD with a dead-zone problem is zero when the input phase difference is very small, and the loop is no longer able to eliminate the phase error. Consequently, the phase noise and the timing jitter of the VCO are increased as depicted in Fig. 3.37.

For PFDs using TSPC D-type flip-flops, the propagation delay can be much smaller and the operation frequency can be much higher than those for traditional PFDs using RS-type flip-flops as shown in Fig. 3.38 (Young *et al.*, 1992). The reason is that the critical reset path consists of six gate delays for conventional PFDs using

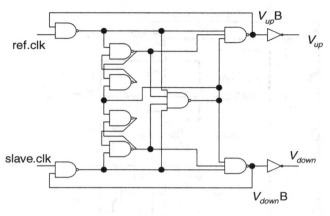

Fig. 3.38 Conventional PFD using RS flip-flops

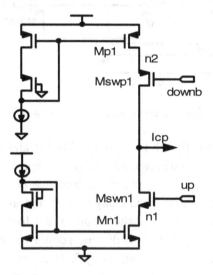

Fig. 3.39 Schematic of a charge pump

RS flip-flops but only three gate delays for PFDs using TSPC D flip-flops (Kim *et al.*, 1997).

3.5. Charge pump

A charge pump (CP) is used to sink and source current into the loop filter based on the outputs of the PFD. A schematic of a CP is shown in Fig. 3.39. Transistors Mn1 and Mp1 provide 'up' and 'down' currents and are combined to generate the required net current to sink from or to source to the loop filter. The switches Mswp1 and Mswn1 are driven by the 'up' and 'downb' output signals of the

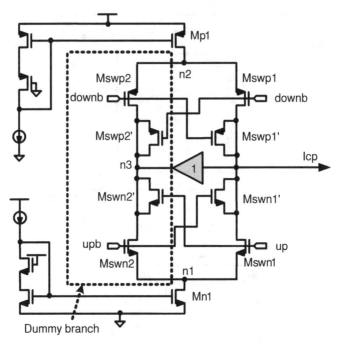

Fig. 3.40 Schematic of charge pump with dummy branch

phase–frequency detector. If both the switches Mswp1 and Mswn1 are turned off, the voltages at nodes n1 and n2 are pulled to ground and V_{DD}, respectively. When the switches are turned on again, charge disturbance is induced by the voltage difference between the output of the charge pump and the voltages at nodes n1 and n2. Ripples, therefore, occur at the output of the loop filter, which result in spurs in the synthesizer output. To remedy this, a dummy branch, which consists of switches Mswp2 and Mswn2, as shown in Fig. 3.40, can be used. Once the switches Mswp1 and Mswn1 are turned off, the switches Mswp2 and Mswn2 are turned on. Consequently, it prevents nodes n1 and n2 from being discharged and charged, respectively. Complementary switches (Mswp1′, Mswp2′, Mswn1′ and Mswn2′) are used to provide complementary charges so as to minimize the error at the loop filter due to charge injection and clock feed-through.

In addition, as shown in Fig. 3.40, if the voltages of node 'out' and node n2 are not the same, a charge sharing problem occurs, which leads to spurious tones at the output of the PLL. Thus, a unity-gain buffer is used to maintain the same voltage potential for the two nodes (Young *et al.*, 1992). The input common mode and the output of the buffer should be able to achieve a large dynamic range in order not to sacrifice the available tuning range of the VCO. Otherwise, a large VCO gain should be designed for the same tuning range requirement. Alternatively, if an active filter is employed as shown in Fig. 3.41, the unity-gain buffer is not necessary. This is

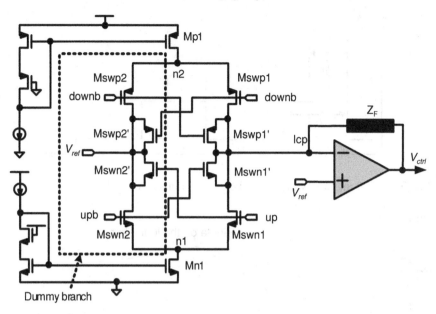

Fig. 3.41 Charge pump associated with an active loop filter

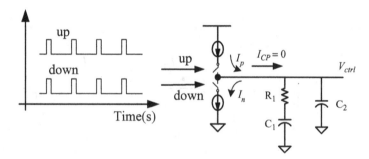

Fig. 3.42 Identical I_p and I_n generating zero average current to the loop filter

because the output node of the charge pump is isolated from the control path of the VCO, which is regulated based on the frequency of interest, while the output voltage of the charge pump does not suffer from output variation as it is biased at a reference voltage V_{ref} in the locked condition.

As mentioned in Section 3.4 regarding the design of a PFD, a minimum delay pulse of the 'up' and 'down' signal is desirable to avoid the dead-zone problem. Therefore, as shown in Fig. 3.42, if the currents I_p and I_n are well matched, the charge pump generates zero average current and zero net charge to the loop filter when the PLL is in its phase-locked condition.

In the steady state, the net charge injected into the loop filter should be zero. If there are mismatches between I_p and I_n, the current pulse width of I_p and I_n should

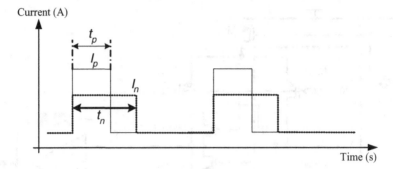

Fig. 3.43 Current pulse waveform when the currents I_p and I_n have mismatches

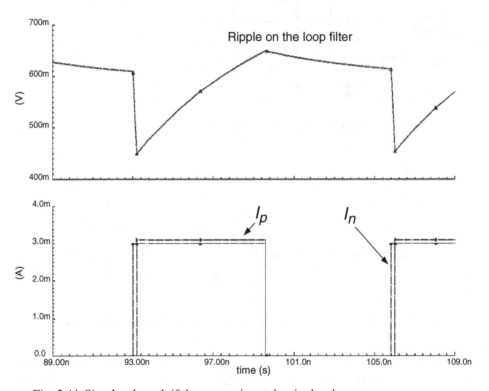

Fig. 3.44 Simulated result if there are mismatches in the charge pump

fulfill the following condition:

$$I_p \cdot t_p = I_n \cdot t_n, \qquad (3.37)$$

and the graphical representation is shown in Fig. 3.43.

Such a phenomenon can generate periodic ripple on the loop filter as well as on the control path of the VCO, as depicted in Fig. 3.44. As a result, large sidebands at a frequency of $\pm f_{ref}/M$ appear at the PLL output. In the time domain, the current

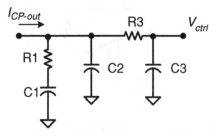

Fig. 3.45 Third-order passive loop filter

mismatches cause a steady phase offset between the two input clocks of the PFD. To alleviate the problem of current mismatch, a common centroid layout with multiple interleaving fingers can help. The sizes of the bias devices can also be made larger.

3.6. Loop filter

A loop filter is often used in PLLs and synthesizers, not only for converting the current from the charge pump to the control voltage for the VCO, but also for filtering out noise coming from the input clock to the control voltage. Otherwise, unacceptably high spurious tones are present in the PLL output spectrum. This section presents different types of loop-filter implementation, including passive filters and active filters.

3.6.1. Third-order passive loop filter

Instead of a second-order loop filter, as shown in Chapter 2, a third-order loop filter (Fig. 3.45) can be used to further suppress ripples at its output, which is also the control voltage of the VCO. With one more pole being added, the transfer function of the loop filter becomes:

$$\frac{V_{ctrl}}{I_{CP-out}} = \frac{k'}{s} \frac{1+s\tau_z}{1+s\tau_{p2}} \cdot \frac{1}{1+s\tau_{p2}} = \frac{1+s(R_1 C_1)}{s(C_1+C_2)(1+sR_1(C_1 \parallel C_2))(1+sR_3 C_3)},$$

(3.38)

where:

k'	is the time constant of integration equal to $1/(C_1 + C_2)$;
τ_z	is the time constant that provides a stabilizing zero to the loop which is equal to $R_1 C_1$;
τ_{p1} and τ_{p2}	are the time constants of the poles that suppress the tones of the reference clock and its higher harmonics. The time constant of τ_{p1} equals $R_1 C_1 C_2/(C_1 + C_2)$, while τ_{p2} equals $R_3 C_3$.

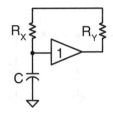

Fig. 3.46 Active loop filter using capacitive multiplication (Larsson, 2001)

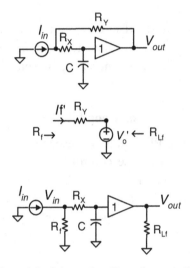

Fig. 3.47 Small signal model of the active filter shown in Fig. 3.46

Although it is simple to design and implement passive loop filters, they are typically too big to be put on-chip.

3.6.2. Active loop filter

The capacitor size used in passive loop filters is typically nanofarads in order to provide sufficient suppression of noise. As a consequence, most of them either occupy a very large chip area or are implemented off-chip. Active filter implementation can help reduce the total capacitance required and make on-chip implementation more feasible. Figure 3.46 shows the schematic of an active loop filter using a capacitive multiplication technique to increase the effective capacitance and minimize the chip area (Larsson, 2001).

The loop filter is actually a shunt–shunt feedback network as redrawn in Fig. 3.47. Using a small-signal model to analyze the network, the parameters R_{LF},

R_f, and the feedback factor f are given as

$$R_{LF} = R_Y,$$
$$R_f = R_Y,$$
$$f = \frac{If'}{v'_o} = \frac{-1}{R_Y}.$$

(3.39)

So, the open-loop gain, A', is

$$A' = \frac{v_{out}}{I_{in}} = \frac{R_Y}{R_Y + \left(R_X + \dfrac{1}{sC}\right)} \cdot \frac{1}{sC}$$

$$= \frac{R_X}{1 + s(R_X + R_Y)C}.$$

(3.40)

The closed-loop gain, A_f, becomes

$$A_f = \frac{A'}{1 + A'f} = \frac{v_{out}}{i_{in}}$$

$$\frac{v_{out}}{i_{in}} = \frac{1}{s\left(\dfrac{R_X + R_Y}{R_Y}\right)C}.$$

(3.41)

If $R_x \gg R_Y$, then $\dfrac{v_{out}}{i_{in}}$ and $\dfrac{v_{in}}{i_{in}}$ are

$$\frac{v_{out}}{i_{in}} = \frac{1}{s\left(\dfrac{R_X}{R_Y}\right)C},$$

$$\frac{v_{in}}{i_{in}} = \frac{1 + sR_YC}{s\left(\dfrac{R_X}{R_Y}\right)C}.$$

(3.42)

The term of $(R_X/R_Y)C$ is the effective capacitance C_{eff}, which has been increased by the ratio of the two resistors R_X and R_Y. This can be maximized to minimize the actual capacitance, C, and its chip area.

Another configuration of active loop filters is the dual-path filter that employs two charge pumps with different currents to create a stabilizing zero without using a real large capacitor (Mijuskovic et al., 1994; Craninckx and Steyaert, 1998; Kan, Leung and Luong, 2002). Such an approach relies on scaling the ratio between the currents of the two charge pumps to control the zero location to minimize the total capacitance required, as opposed to scaling a ratio of two resistors as discussed

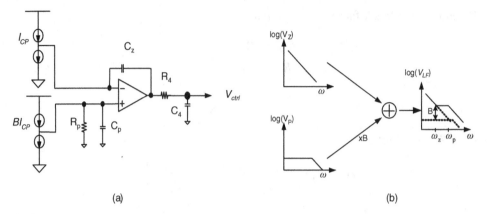

Fig. 3.48 (a) Dual-path filter implementation, (b) frequency response of the dual-path filter

above. Graphically, the stabilizing zero can be realized by combining the outputs of the two charge pumps with a current ratio of B as illustrated in Fig. 3.48. However, increasing the current ratio B results in an increase in the current noise from the charge pump.

In addition to minimizing the physical size of the on-chip capacitors, the dual-path filter can also achieve a large voltage range for the control voltage of the VCO. Moreover, the filter can provide a fixed reference voltage to the output of the charge pumps, which helps prevent the charge pumps from suffering a large output swing and thus minimizes the mismatch between the sink and source currents of the charge pumps. The circuit implementation of the dual-path filter in a synthesizer design will be discussed in detail in Chapters 7 and 8.

3.7. Inductor design

Inductors are key elements in designing high-frequency integrated circuits and have been extensively used in both radio frequency integrated circuits and clock recovery circuits. Traditionally, the use of off-chip inductors was dominant because of their high quality factor Q. However, recently, different types of inductors, including bond wires and on-chip spiral inductors, have been extensively investigated and demonstrated for integrated circuits. Bond wires commonly used for connection between chip and package have been proposed to replace low-Q on-chip inductors to obtain better performance because their quality factor Q can be as high as 40 (Steyaert and Craninckx, 1994). This can definitely help in reducing power consumption and improving the phase noise of VCOs. Nonetheless, bond wires have big problems with accurate modelling, and with their reliability and repeatability, which make them less attractive.

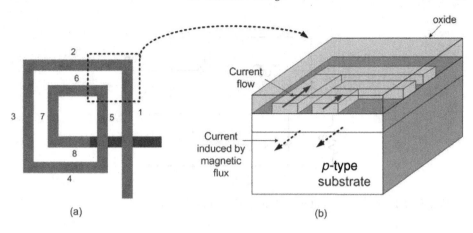

Fig. 3.49 (a) Top view and (b) cross section of a two-turn spiral inductor

Compared with off-chip inductors and bond wires, on-chip spiral inductors are much more desirable because they can enable single-chip integration. The reproducibility from chip to chip using an on-chip inductor is better than utilizing bond wires. However, the major problem with on-chip inductors is their high loss or low quality factor. Modern CMOS technology with an epitaxial layer makes use of a heavily doped substrate (approximately 0.01 Ω-cm) to minimize the latch-up problem. Nevertheless, the magnetic field generated by on-chip inductors induces eddy currents in the substrate and increases the loss to the substrate. Etching out the substrate underneath the spiral coil can alleviate the loss problem (Chang, Abidi and Gaitan, 1993), but it requires post-processing steps which unavoidably increase production cost. Mixed-signal RF processes have become available recently with less conductive substrates (approxmately 10 Ω-cm) and a thick top metal layer to improve performance in terms of quality factor and resonant frequency. In contrast, less conductive substrates may introduce a potential problem of latch-up, but this problem can be avoided by careful layout with sufficient guard rings.

3.7.1. Fundamentals of on-chip inductors

Figure 3.49 shows the typical top view and cross section of a spiral inductor that utilizes the topmost metal layer available for maximum Q and the next top metal layer passes under to provide a connection to the terminal at the center. The design parameters of an inductor are given by the number of turns (n), the metal width (w), the metal-to-metal spacing (s), the inner-hole diameter (d_{in}), and the outer diameter (d_{out}). In fact, the physical size of on-chip inductors is relatively small compared with the wavelength of operation and, as a result, it is quite accurate to use lumped

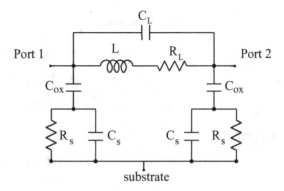

Fig. 3.50 Lumped-element model of an on-chip inductor

models with passive elements to characterize on-chip inductors for analysis and simulations. A typical lumped-element model for on-chip inductors is illustrated in Fig. 3.50 (Long and Copeland, 1997; Niknejad, Gharpurey and Meyer, 1998). The π-equivalent model consists of nine passive elements including:

L the total inductance produce by the metal winding and mutual coupling;
R_L the total loss of the inductor;
C_{ox} the capacitance between inductor and substrate;
R_s and C_s the substrate parasitic effects.

The spiral inductance is composed of two components: self-inductance and mutual inductance (Greenhouse, 1974). The total inductance of the coil, L_{tot}, can be expressed as

$$L_{tot} = L_{self} + M_+ - M_- \tag{3.43}$$

where:

L_{self} is the total self-inductance of all conductor segments;
M_+ is the mutual inductance due to positive coupling between conductors;
M_- is the mutual inductance due to negative coupling between conductors.

As illustrated in Fig. 3.49, assuming that the total number of segments is 8, the self-inductance of the spiral inductor is given by

$$L_{self} = \sum_{i=1}^{8} L_i$$

$$L_i = 0.002 \cdot l_i \cdot \left\{ \ln \left[\frac{2l_i}{w+t} \right] + 0.500\,49 + \frac{w+t}{3l_i} \right\}, \tag{3.44}$$

where:

L_i is the DC self-inductance (in nanohenries);
l_i is the length of the ith segment (in centimetres);
w is the width of metal segment (in centimetres);
t is the metal thickness (in centimetres).

The mutual inductor is formed by the magnetic coupling between conductors and can be positive or negative depending on whether the currents in the two conductors have the same (in-phase) or opposite (out-of-phase) flowing directions, respectively. Thus, M_+ and M_- are given by

$$M_+ = 2(M_{15} + M_{26} + M_{37} + M_{48}) \tag{3.45}$$

$$M_- = 2(M_{13} + M_{17} + M_{24} + M_{28} + M_{53} + M_{57} + M_{64} + M_{68}), \tag{3.46}$$

where M_{ij} is the coupling between conductor i and conductor j, which can be estimated as

$$M_{ij} = 2 \cdot I \cdot Q_{ij}. \tag{3.47}$$

Q_{ij} is the mutual-inductance parameter which is given as (Greenhouse, 1974)

$$Q_{ij} = \ln \left[\frac{1}{\text{GMD}} + \sqrt{1 + \frac{I^2}{\text{GMD}^2}} \right] - \sqrt{1 + \frac{\text{GMD}^2}{I^2}} + \frac{\text{GMD}}{I}, \tag{3.48}$$

where GMD is the geometric mean distance between two conductors, for example, segment 1 and segment 3, which is roughly equal to the distance between the center-to-center of the two conductors. A more accurate expression is given as (Greenhouse, 1974)

$$\ln \text{GMD} = \ln d - \frac{w^2}{12d^2} - \frac{w^4}{60d^4} - \frac{w^6}{168d^6} - \frac{w^8}{360d^8} - \frac{w^{10}}{660d^{10}} - \cdots. \tag{3.49}$$

From the above equations, it can easily be concluded that the inductance is increased by using longer segments and a larger number of turns. A larger number of turns can increase the mutual inductance without the need of longer conductors. It follows that the DC resistance is decreased and the quality factor is improved. In addition, circular and octagonal inductors generally show roughly 10% less resistance and higher Q than square inductors (Chaki et al., 1995). However, these non-standard inductors are built out of many small segments and thus take much more time to analyze and simulate. Besides, it should be noted that the inner hole of spiral inductors is usually maximized as this area contributes minimum inductance but high loss.

Fig. 3.51 Inductor with loss modeled by a series resistor

3.7.2. *Quality factor (Q) of on-chip inductors*

In addition to its inductance value and resonant frequency, the quality factor, Q, is another critical parameter of on-chip inductors that requires much attention and optimization. The quality factor is defined as the ratio between energy stored and the energy loss, and is given by

$$Q = 2\pi \frac{\text{energy stored}}{\text{energy lost per cycle}}. \tag{3.50}$$

Inductors can store and release magnetic energy in every cycle, and Equation (3.51) can be used to quantify how much energy is stored by an inductor with an inductance value L:

$$E_{\mathrm{L}} = \frac{L \cdot i_{\mathrm{L}}^2}{2}, \tag{3.51}$$

where i_L is the current flowing through the inductor.

As illustrated in Fig. 3.51, the loss in inductors can be simply modeled by a series resistor R_{L}, and the quality factor, Q_{L} of inductors is given by

$$Q = 2\pi \frac{\text{peak magnetic energy stored}}{\text{energy lost per cycle}} \tag{3.52}$$

and

$$Q_{\mathrm{L}} = 2\pi \frac{\dfrac{L \cdot |i_{\mathrm{L}}|^2}{2}}{\dfrac{R_{\mathrm{L}}|i_{\mathrm{L}}|^2}{2T}}$$

$$= \frac{\omega L}{R_{\mathrm{L}}}. \tag{3.53}$$

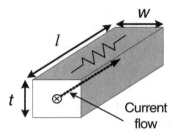

Fig. 3.52 Conductor with loss

where $|i_L|$ is the peak current passing through the inductor, T is the period of a cycle, and ω is the operating frequency.

It is important to note that in modern CMOS processes, the quality factor of capacitors can generally be higher than 20 while that for on-chip inductors is limited to around 5 to 10 in the gigahertz frequency range. This is the reason why the performance of LC oscillators is mainly dominated and limited by the inductors' quality factors. In addition, because the quality factors of inductors are inversely proportional to their loss but directly proportional to their inductance value, it is desirable to maximize the inductance value while minimizing the loss.

Resistance in the conductor opposes the current flow and dissipates energy as loss. In general, the resistance of a conductor (R), as shown in Fig. 3.52 is given by

$$R = \frac{\rho \cdot l}{w \cdot t},\qquad(3.54)$$

where:

ρ is the resistivity of the conductor (in Ω-m);
l is the total length of the conductor (in metres);
w is the width of the conductor (in metres);
t is the thickness of the conductor (in metres).

However, the foregoing equation is only valid at DC operation. At high frequencies, current density in conductors becomes non-uniform due to the skin effect and the proximity effect (Yue and Wong, 2000). Skin effect is a common phenomenon at high frequencies in which the electromagnetic field and the current in the conductor decay rapidly and are confined to flow near the surface of the conductor, which effectively increases the resistive loss and reduces the quality factor of the conductor. The proximity effect is another high-frequency phenomenon due to eddy currents induced among conductors, which adversely affect the property of metal. To account for these effects, a more accurate expression for R_L can be represented by

$$R_{\mathrm{L}} = \frac{l}{w \cdot \sigma \cdot \delta (1 - e^{-l/\delta})}, \tag{3.55}$$

where:

- σ is the conductivity of the conductor;
- δ is skin depth which is given by

$$\delta = \sqrt{\frac{1}{\pi \sigma \mu f}}, \tag{3.56}$$

where:

- μ is the permeability of free space, μ_o (equal to $4\pi \times 10^{-7}$ H/m);
- f is frequency (in hertz).

In addition to the conductor loss, other parasitics also dissipate the energy stored by the inductor such as C_{L}, C_{ox}, C_{si} and R_{si}. C_{L} is the overlapping capacitance between the metal layers of the spiral inductor and can be approximated by

$$C_{\mathrm{L}} = n \cdot w^2 \cdot \frac{\varepsilon_{ox}}{t_{oxm}}, \tag{3.57}$$

where:

- n is the number of turns;
- w is the metal width;
- ε_{ox} is the permittivity of the oxide layer between the two conductors;
- t_{oxm} is the thickness of the oxide layer between the two conductors.

The capacitance between the spiral and the substrate, C_{ox} is expressed as

$$C_{ox} = \frac{1}{2} w \cdot l \cdot \frac{\varepsilon_{ox}}{t_{ox}}, \tag{3.58}$$

where t_{ox} is the thickness of the oxide layer between the inductor and the substrate. The parasitics, R_{si} and C_{si}, used to model the substrate loss are given by

$$R_{si} = \frac{2}{w \cdot l \cdot G_{sub}}, \tag{3.59}$$

$$C_{si} = \frac{w \cdot l \cdot C_{sub}}{2}. \tag{3.60}$$

G_{sub} is the substrate conductance per unit area, while C_{sub} is the substrate capacitance per unit area.

The foregoing model, in Fig. 3.50, only considers the electric coupling between the inductor and the substrate. At high frequencies, magnetic coupling should be

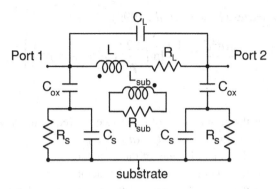

Fig. 3.53 Lumped-element model of inductor including magnetic coupling from substrate

taken into account. Lenz's law indicates that eddy current is generated under the influence of magnetic flux, which tends to oppose the original field. The resultant inductance of the on-chip inductor is therefore reduced while the series resistance is effectively increased. Moreover, these changes are frequency dependent. To take this phenomenon into account and achieve a more accurate model and simulation, two extra elements, inductance L_{sub} and resistance R_{sub}, can be included in the π-equivalent model as illustrated in Fig. 3.53. The two elements are formed as a secondary transformer. Hence, the resultant impedance of the on-chip inductor can be approximated as (Zheng *et al.*, 2000):

$$Z_{ind}(\omega) \approx R_{\mathrm{L}} + j\omega L - \frac{M(j\omega)}{R_{sub} + j\omega L_{sub}}. \tag{3.61}$$

Therefore, the resultant inductance L' can be expressed as

$$L' \approx L - kL_{sub} \tag{3.62}$$

and the resultant series resistance R_{L}' becomes

$$R' \approx R_{\mathrm{L}} + kR_{sub}, \tag{3.63}$$

where k is given by

$$k = \frac{\omega^2 M}{R_{sub}^2 + \omega^2 L_{sub}^2}. \tag{3.64}$$

3.7.3. Design guidelines for on-chip inductors

On-chip inductors can be designed based on the following equation, which is based on an empirical model-fitting process to estimate the inductance value for inductors with different geometries (Mohan *et al.*, 1999).

$$L = \frac{\mu n^2 d_{avg} c1}{2} \left[\ln \left(\frac{c2}{\rho} + c3\rho + c4\rho^2 \right) \right] \tag{3.65}$$

where:

μ	is the permeability of free space, μ_o (equal to $4\pi \times 10^{-7}$ H/m);
ρ	is the ratio $(d_{out} - d_{in})/(d_{out} + d_{in})$;
d_{in}	is the inner diameter of the inductor;
d_{out}	is the outer diameter of the inductor;
d_{avg}	is the average diameter of the inductors equal to $(d_{out} + d_{in})/2$;
$c1, c2, c3$ and $c4$	are model-fitting parameters depending on the geometrical layout (Mohan *et al.*, 1999).

In order to estimate more accurately the inductance value and to obtain some rough and relative prediction of the quality factor Q, CAD simulation tools such as ASITIC (Niknejad and Meyer, 1998b), FastHenry (Kamon, Tsuk and White, 1994), Sonnet, or Momentum (from ADS) may be used.

Among these simulation tools, ASITIC, FastHenry and Momentum are found to offer fast simulation speed, while ASITIC is somewhat more user-friendly. In particular, given technology parameters, ASITIC can simulate inductance, quality factor, resonant frequency, coupling factor between two inductors, and most of the model parameters, with some reasonable accuracy. In addition, it can also automatically generate inductors with different shapes, such as straight, square octagonal, and circular ones.

While all these simulation tools are found to be able to accurately predict the inductance value, they still cannot model and estimate the loss and the quality factor accurately. In general, the quality factor is overestimated, but the estimate can still be helpful as a relative measure for Q optimization and becomes quite useful for Q prediction after a correction factor found from measurements is incorporated. To achieve more accurate results, an EM simulator such as Sonnet can be used but the simulation time is generally too long to be useful in practice.

3.8. Varactor design

A varactor is a voltage-dependent variable capacitor commonly used to tune the oscillation frequency of both LC and ring VCOs by changing the control voltage across the varactor and thus its capacitance. The most critical parameter in designing varactors is their tuning ratio, defined as the ratio between the maximum and minimum capacitance values, C_{max}/C_{min} they can provide. As for inductors, another important design parameter for varactors is their quality factor Q. Although the overall quality factor of resonant LC tanks is typically dominated by the inductor Q, as mentioned earlier, careful attention needs to be paid to design varactors with maximum Q, or else their Q may be low enough to become significant in degrading the overall Q and thus the overall performance of LC oscillators in terms of phase noise and power consumption. In the following sections, after a brief review of capacitor Q, design considerations and trade-offs of the *pn*-junction varactor and the accumulation-mode varactor are described and compared.

3.8.1. *Quality factor (Q) of capacitors*

As reviewed in Section 3.7.2, inductors store and release magnetic energy. Similarly, electric energy is stored and released in capacitors in every oscillation cycle. The electric energy is given in Equation (3.66).

$$E_c = \frac{C \cdot v_C^2}{2},\qquad(3.66)$$

where v_C is the voltage across a capacitor with a capacitance value C.

Although typically much higher than that of on-chip inductors, the quality factor of capacitors used in LC oscillators may become low enough to become detrimental if not optimized properly. The quality factor of a capacitor (Q_C) is given by

$$Q = 2\pi \frac{\text{peak electric energy stored}}{\text{energy lost per cycle}}\qquad(3.67)$$

and

$$\begin{aligned}
Q_C &= 2\pi \frac{\dfrac{C \cdot |v_C|^2}{2}}{\dfrac{R_C |i_C|^2}{2T}} \\[2mm]
&= 2\pi \frac{\dfrac{|i_C|^2}{2\omega^2 C}}{\dfrac{R_C |i_C|^2}{2T}} \qquad(3.68)\\[2mm]
&= \frac{1}{\omega C R_C},
\end{aligned}$$

where v_C is the peak voltage across the capacitor and i_C is the peak current passing through the capacitor as illustrated in Fig. 3.54.

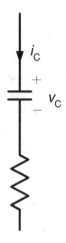

Fig. 3.54 Capacitor with loss modeled by a series resistor

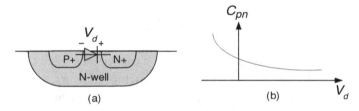

Fig. 3.55 (a) Cross section of the *pn*-junction varactor, (b) capacitance against
voltage

3.8.2. pn-*junction varactors*

A *pn*-junction varactor can be simply realized by using p_+ diffusion in an *n*-well as
shown in Fig. 3.55(a). The junction capacitance between p_+ active and the *n*-well
is depletion capacitance and voltage dependent.

By changing the voltage potential across the varactor, the effective capacitance
can be tuned as depicted in Fig. 3.55(b). The variation of capacitance is relatively
linear over the positive region of the control voltage. However, one critical limitation of *pn*-junction varactors is that their control voltage should be limited so
as not to forward bias the junctions. Otherwise, the junctions will be turned on,
and their quality factor will drop significantly. Ironically, *pn*-junction varactors
achieve their maximum tuning capacitance ratio when they are forward-biased as
illustrated in Fig. 3.55(b) (Porret *et al.*, 2000). For this reason, *pn*-junction varactors cannot be optimized to achieve both a large tuning range and a high quality
factor.

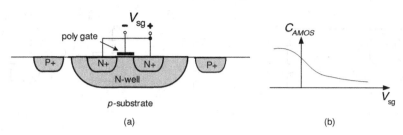

Fig. 3.56 (a) Cross section of accumulation-mode varactor, (b) capacitance against voltage

3.8.3. *Accumulation-mode varactors*

Figures 3.56(a) shows a cross section of an accumulation-mode varactor. The structure is similar to that of a NMOS transistor but with the NMOS device being placed in an *n*-well instead of a *p*-well. In addition, the drain and the source terminals are connected together. The capacitance is formed between the gate and the substrate and is varied with the control voltage V_{sg} being applied across the gate and the drain–source terminal.

When V_{sg} is negative, the voltage potential at the gate of the varactor is larger than that at the source. The positive charge in the gate causes accumulation in the *n*-well and leads to an increase in the concentration of majority carriers near the surface. As a result, the device becomes more conductive; the capacitance is mainly contributed by the oxide capacitance and becomes larger.

When V_{sg} is increased, the gate becomes more negative, and negative charge in the *n*-well region under the gate becomes depleted. The depletion region gets wider and the depletion capacitance gets smaller as V_{sg} is increased. As this depletion capacitance is equivalently connected in series with the oxide capacitance, the effective overall capacitance becomes smaller. This phenomenon is illustrated in Fig. 3.56(b).

Because the majority carrier type is the electron, which intrinsically has higher mobility, the quality factor of such accumulation-mode varactors can be higher than their inversion-mode counterparts (Andreani and Mattisson, 2000). To achieve even better Q, larger number of fingers should be employed, and the gate length should be minimized. On the other hand, compared with *pn*-junction varactors, accumulation-mode varactors are more non-linear for the same voltage variation. The non-linearity of the varactors in turn introduces non-linearity in the VCO gain, which affects and limits the stability and the phase-noise optimization of the PLL system. The *pn*-junction varactor has better linearity and the quality factor is also relatively close to that of accumulation-mode varactors (Andreani and Mattisson, 2000). Nevertheless, the tuning capacitance ratio is smaller, as mentioned.

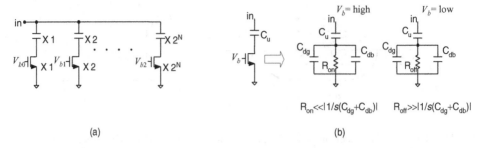

(a) (b)

Fig. 3.57 (a) Schematic of switched-capacitor array, (b) model of an SCA unit

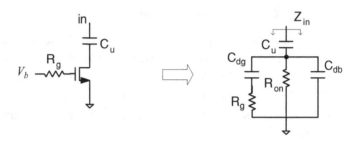

Fig. 3.58 SCA implementation with resistors in series with the gate

3.9. Switched-capacitor array (SCA)

A switched-capacitor array (SCA) can be used to provide effectively a coarse tuning of the total capacitance and thus of the oscillation frequency of an oscillator. Basically, an SCA is constructed with several switched-capacitor branches in parallel, each of which consists of a unit capacitor in series with a MOS transistor as a digital switch. By digitally controlling the gates of the MOS switches, the total capacitance can be controlled, and the oscillation frequency can be tuned. The unit capacitors in an SCA can be designed to be of the same value to achieve good matching and uniformity or to be binary-weighted to minimize the number of branches required. Figure 3.57 shows the schematic of a binary-weighted switched-capacitor array.

The MOS switches in an SCA are not perfect. Finite turn-on resistance (R_{on}) limits the quality factor of the capacitor. In addition, non-zero drain-to-bulk capacitance (C_{db}) and drain-to-gate capacitance (C_{dg}) are connected in series with the unit capacitor (C_u) and presents a finite parasitic capacitance when the switches are turned off, which inevitably degrades the maximum achievable tuning range for the SCA. A smaller turn-on resistance and thus a larger Q can be realized by using a larger switch size. Nevertheless, the parasitic capacitance will be increased, and the tuning range becomes smaller. In addition, the effect of the drain-to-source capacitance can be minimized by connecting a resistor between the gate and the control voltage, as illustrated in Fig. 3.58. The impedance of the SCA connected with resistors in series with the gate is given by

$$Z_{in} = \frac{1}{sC_u} + \frac{1}{\dfrac{sC_{dg}}{1 + sC_{dg}R_g} + \dfrac{1}{R_{on}} + sC_{db}}. \tag{3.69}$$

However, the minimum parasitic capacitance of the SCA with the switches off is still limited by the drain-to-bulk capacitance. For given aspect ratios, transistors in the donut shape can be used to minimize their parasitic capacitance and thus increase the tuning range of the SCA without sacrificing its quality factor Q (Kral, Behbahani and Abidi, 1998). In any case, there still remains a trade-off between the quality factor and the capacitance tuning range in designing a SCA as a varactor.

4

Low-voltage design considerations and techniques

4.1. Introduction

The impact of low-voltage design is degradation in speed, because of limited driving capability, and in signal-to-noise ratio (SNR), because of the reduced signal swing. In addition, for synthesizer designs, a low supply voltage reduces the frequency tuning range and degrades the phase noise unless the current and power consumption are increased. Moreover, the design of prescalers and high-speed digital circuits becomes much more challenging because of the speed degradation of digital circuits with a low-voltage supply. This chapter discusses these design considerations and presents some of the design techniques required for critical building blocks in RF CMOS synthesizers as the supply voltage is lowered.

4.2. System considerations

The control voltage of the VCO in a synthesizer becomes limited under a low supply voltage. This results in a limited frequency tuning range with a given VCO gain. A larger VCO gain could be used to compensate for the degradation of the tuning range at the expense of the phase-noise and spurious-tone performance of the system. A high loop bandwidth helps to improve rejection of the VCO phase noise but sacrifices the spurious-tone suppression. In contrast, a small PLL loop bandwidth can provide larger spurious-tone suppression but results in less rejection of the VCO phase noise. On other hand, LC oscillators can achieve better phase noise than ring oscillators for a given power. As a consequence, a synthesizer using an LC oscillator with a small loop bandwidth can attain optimized performance in terms of noise and spurious suppression. However, the disadvantages of using an LC oscillator are its limited tuning range and its large chip area.

Furthermore, as the supply voltage is lowered, noise and voltage ripples on the supply line become more significant, and some building blocks in the loop may fail

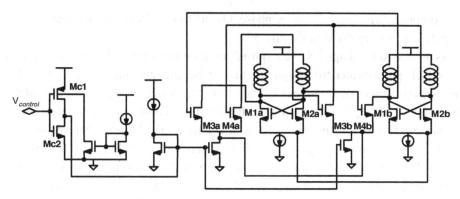

Fig. 4.1 Schematic of a quadrature LC VCO (Leung and Luong, 2003)

to function properly or adequately. Large decoupling capacitors can be connected to the supply line to help suppress the noise and ripples. However, the technique is not effective at low frequencies unless extremely large capacitance and chip area are used.

4.3. Voltage-controlled oscillators

When the supply voltage is scaled down, the available voltage range for varying capacitance in an LC VCO is also reduced. It would be more desirable to focus on maximizing the capacitance tuning range for a given tuning range of the control voltage. *pn*-junction varactors may become unattractive because of their small capacitance tuning ratios. Accumulation-mode varactors, which have higher capacitance tuning ratios, should be employed for fine tuning of the total capacitance. Moreover, a switched-capacitor array (SCA) should be included for coarse tuning, not only to compensate for process variation, but also to reduce the frequency range required for fine tuning. The latter is quite useful in reducing the VCO gain to improve the overall performance as previously emphasized. In addition, if both in-phase and quadrature-phase outputs are required, coupled LC oscillators can be used, and an effective mechanism to achieve fine frequency tuning is available by tuning the current coupling between the LC tanks. Such tuning becomes quite essential in low-voltage VCO designs because varying the current coupling, unlike in varactors, is not limited by the supply voltage. Power consumption as well as chip area, therefore, can be saved without degrading the noise performance of the VCO significantly.

Figure 4.1 shows a proposed quadrature LC/VCO (Leung and Luong, 2003) that fully utilizes the available voltage range of the output of the loop filter. The circuit comprises two identical LC oscillators, which are coupled to each other and have their outputs oscillating 90° out of phase. Owing to its differential architecture, the outputs have a duty cycle of exactly 50%.

As discussed in Chapter 3, the required transconductance of the negative-g_m cell is inversely proportional to the square root of inductance and directly proportional to the series resistance of the inductor. Therefore, the capacitance at the outputs of the quadrature VCO (QVCO) should be minimized to minimize power consumption. If a varactor is used for frequency tuning, it will need to be large to achieve the same tuning ratio at a low supply voltage, and the inductor would need to be reduced further. This will, in turn, require larger transconductance and power consumption.

The frequency of the proposed QVCO shown in Fig. 4.1 is not tuned by *pn*-junctions or MOS capacitors. Instead, frequency tuning is based on varying the tranconductances of the coupling transistors M_{3a-3b} and M_{4a-4b}. To achieve a low VCO gain (K_{VCO}), M_{c1} and M_{c2} are inserted to achieve rail-to-rail frequency tuning. However, when the control voltage is in the middle of the supply, the VCO gain becomes non-linear because the two transistors operate in the weak inversion. In order to compensate for this non-linearity, the threshold voltage of the PMOS device is reduced to ensure the device is in strong inversion at the middle of the supply. The threshold voltage of a MOS transistor as a function of the bulk-source voltage is shown in Equation (4.1):

$$V_{th} = V_{th0} + \gamma(\sqrt{|2\phi_F - V_{BS}|} - \sqrt{|2\phi_F|}, \tag{4.1}$$

where V_{th0} is the threshold voltage when $V_{BS} = 0$, γ is the body-effect coefficient, and $2|\phi_F|$ is the surface potential at strong inversion.

By biasing the bulk at a voltage larger than the source voltage, the threshold voltage can be decreased. However, this biasing scheme may result in a large amount of leakage current to the substrate. To overcome the problem, the current-driven-bulk (CDB) technique is proposed to turn on the *pn*-junction from the source to the bulk for the PMOS with limited current flow (Lehmann and Cassia, 2001). With the CDB technique, the threshold voltage can be reduced, and the PMOS transistor can work in the strong inversion in the voltage range around half of V_{DD}. Measurement shows that the K_{VCO} is very linear over the whole output voltage range of the active loop filter.

4.4. Divide-by-2 circuit

The speed degradation under low-voltage supply is a bottleneck to designing high-frequency digital circuits. As the supply voltage is decreased, the gate-to-source driving voltage is also reduced, which leads to an increase of the output RC time constant and of prolongation of the circuit delay. Hence, digital circuits are not

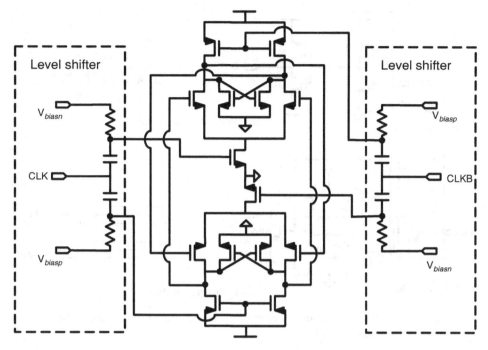

Fig. 4.2 A divide-by-2 with level-shifters for operating at low supply voltage

suitable for use in high-frequency dividers at low supply voltages. In contrast, analog dividers, like SCL divide-by-2 circuits, can have superior performance for high-frequency operation because, unlike in digital circuits, their devices operate in the saturation region for most of the time. Nevertheless, a low-voltage supply still presents a big challenge to the design of analog dividers at high frequencies. As the speed is mainly dependent on the ratio between the transconductance and the capacitive load, limited driving voltage leads to limited transconductance and the limited operation frequency of the dividers. Larger device size can be used to compensate for the reduction of the transconductance at low supply voltages, but this results in larger capacitance, and in the end the frequency still becomes limited.

An SCL divide-by-2 circuit with dynamic loading has advantages in terms of operating frequency compared with SCL using resistive loads. Although the speed is not as high as that of injection-locked dividers using inductive loads, its wider frequency range is an attractive solution to overcoming the process variation. Figure 4.2 shows a divide-by-2 circuit with dynamic loading operating at a low supply voltage. It is similar to the circuit described in Section 3.2.2 except that level shifters are included at the inputs, not only to couple the high-frequency input clock

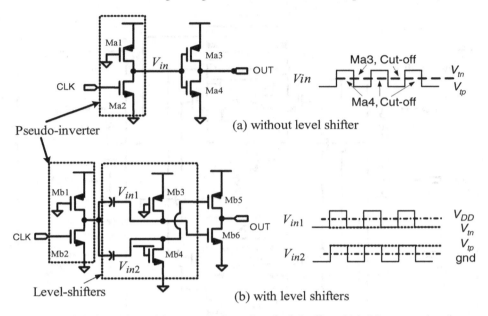

(a) without level shifter

(b) with level shifters

Fig. 4.3 Schematic and input waveform for clock buffers (a) with conventional inverters, (b) with level shifters

signals, but also to bias properly the loading transistors and the current sources. This configuration can work well below 1 V without being limited by the common mode voltage of the previous VCO stage.

4.5. High-speed clock buffer

Although true single-phase clock (TSPC) dividers are simple and fast compared with static-logic dividers, they require rail-to-rail input signals for high-frequency operation. If the input signal swing is not large enough, TSPC dividers will fail to function properly. Therefore, clock buffers are needed to provide rail-to-rail signals to TSPC dividers. As illustrated in Fig. 4.3(a), clock buffers using conventional inverters have a speed limitation because the input swing to M_{a3-4} is bounded within the supply and the transistors, operating in the cut-off region over half of the input period, take a long time to become activated. To fully utilize the input swing to maximize the speed, level shifters are used to shift up or down the output swing from the pseudo-inverter as shown in Fig. 4.3(b). AC-coupling capacitors and pull-up and pull-down devices, Mb3 and Mb4, are proposed to shift the input V_{in1} of NMOS Mb6 close to V_{DD} and input V_{in2} of PMOS Mb5 close to zero to prevent the transistors Mb5 and Mb6 from being turned off completely. As a consequence, the input signal to M_{b5-6} is pushed to exceed the rail-to-rail supply voltages to achieve faster speed and triggering. It enables a larger input voltage drive to the TSPC divider at high frequency under low voltage compared with that of conventional designs using inverter chains with the same device sizes. The pseudo-inverter is needed to isolate and to reduce the actual input loading of the clock buffers. A small conventional inverter could also be used at even lower power consumption but the operation frequency would be much lower. The disadvantage of using this approach is the large chip area needed for the capacitors.

4.6. Prescaler design

As mentioned in Chapter 3, dual-modulus prescalers can be used to implement pulse-swallow frequency dividers and multi-modulus prescalers. However, the main design problem with such a dual-modulus prescaler is that it needs to toggle its two division ratios at the same speed as that of the input clock signal. It follows that they are not suitable for high-frequency prescaler design, in particular at a low supply voltage. A phase-switching or phase-selection approach can relax the speed requirement of the divider circuits because the phase switching takes place at much lower frequencies without sacrificing the division step. To eliminate a potential glitch problem occurring during phase switching, a backward phase-selection scheme, which works well at low voltages, can be combined with a glitch-free phase-selection circuit. This is described in detail in Chapter 8.

4.7. Charge pump

As mentioned previously, the VCO gain needs to be increased as the supply voltage is reduced to maintain the same frequency tuning range. Such voltage-supply reduction also has a detrimental effect on charge pump (CP) design. Cascode current sources are typically used in CPs to provide a high output impedance, as shown in Fig. 4.4, but they may become inapplicable without enough voltage headroom. This will further narrow the output dynamic range of the CP because there is some voltage drop across each device, thereby limiting the tuning range of the VCO.

By implementing the CP together with an active loop filter, as shown in Fig. 3.41, the output node of the CP is isolated from the control path of the VCO. The tuning range of the VCO, as a result, depends on the output stage of the op amp in the active loop filter rather than the output of the CP. If the input stage of the op amp is implemented by PMOS, as illustrated in Fig. 4.5, the output node voltage of the CP is given by Equation (4.2):

$$V_{CP\text{-}out} = V_{DD} - V_{sd,M2} - |V_{tp}|, \tag{4.2}$$

where V_{sd} is the source-to-drain voltage, and V_{tp} is the threshold voltage of the PMOS device.

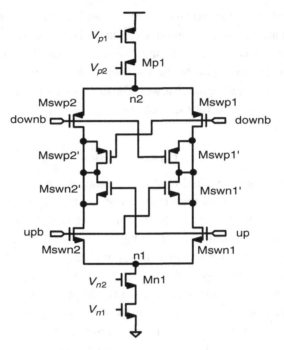

Fig. 4.4 Charge pump with cascode current source

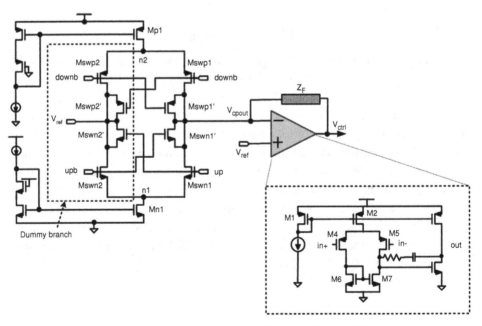

Fig. 4.5 Implementation of charge pumps with dual-path loop filter

The equation indicates that cascode implementation of CPs is not feasible with a low-voltage supply. Fortunately, a dual-path loop filter can be employed to enable the CPs to have limited output swing without sacrificing the output voltage range of the loop filter. Moreover, such a constant output voltage of the CPs is helpful in minimizing the current mismatch of the CPs and thus minimizing spurious tones at the VCO output.

5

Behavioral simulation

5.1. Introduction

Performing transistor-level transient simulation of frequency synthesizers normally takes a very long time. As a result, it is typically neither practical nor worthwhile to verify the transient performance of the synthesizers with simulations at the transistor level. Alternatively, carrying out system-level simulations or behavioral simulations can be quite important and critical in optimizing design parameters and in boosting up the efficiency and verification of a design. A suggested design procedure for PLLs is shown in Fig. 5.1. The first step is the specification definition, in which all the critical parameters are determined, including architecture, output frequency, input frequency, division ratio N, VCO gain K_{VCO}, charge pump current I_{CP}, and loop bandwidth BW. After defining the specification, mathematical models and behavioral models can be constructed for each building block, and for the whole synthesizer, to verify the stability. The issues and considerations of system simulations of RF frequency synthesizers are discussed in this chapter.

5.2. Linear model

The block diagram of a third-order synthesizer is shown in Fig. 5.2. The critical parameters to be defined first are the division ratios N_1 and N_2 as the synthesizer system. The division ratios are determined based on the relation between the input reference and the output frequencies. If a desired output clock frequency is to be 3.2 GHz, while the input clock frequency is 100 MHz, this leads to the division ratios N_1 and N_2 to be 1 and 32, respectively.

After the division ratios have been selected, it is important to determine the required VCO gain, which is strongly dependent on the required frequency tuning range and the maximum swing of the VCO control voltage. The tuning range of a VCO can be achieved by both fine tuning and coarse tuning. Coarse tuning using

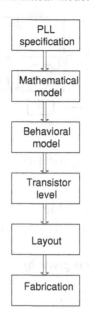

Fig. 5.1 Design flow of phase-locked loops

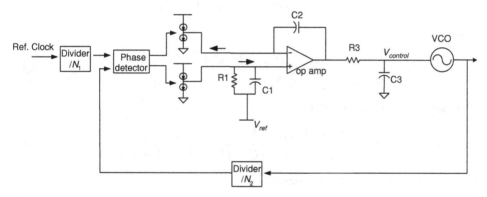

Fig. 5.2 Block diagram of the synthesizer example

SCAs is typically controlled by external digital signals unless an on-chip high-performance analog-to-digital converter is included, which is highly unlikely. On the other hand, fine tuning definitely needs to be controlled by the loop itself to achieve a phase-locked condition. As a result, fine tuning is used to define the VCO gain when analyzing the stability and the required loop bandwidth of the system. Assuming the VCO needs to be tuned by 200 MHz over 80% of the supply voltage of 1 V, the minimum VCO gain is

$$K_{VCO} = 2\pi \frac{200 \text{ MHz}}{0.8 \cdot 1 \text{ V}} = 2\pi \cdot 250 \text{ MHz/V}. \tag{5.1}$$

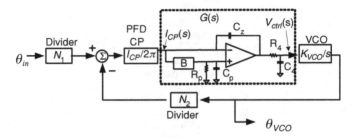

Fig. 5.3 Linear model of PLL using dual-path loop filter

The linear model of the PLL is depicted in Fig. 5.3 and can be used to calculate the component parameters and to analyze the loop performance when the PLL is in the locked condition.

The loop filter transfer function is given by:

$$G(s) = \frac{V_{ctrl}(s)}{I_{CP}(s)}$$

$$= \frac{1}{sC_2} \cdot \frac{1 + s\tau_Z}{1 + s\tau_p} \cdot \frac{1}{1 + s\tau_4}, \tag{5.2}$$

where

$$\tau_Z = R_p(C_p + B \cdot C_z)$$
$$\tau_p = C_p R_p \quad \text{and} \quad \tau_4 = C_4 R_4. \tag{5.3}$$

The loop gain, $A(s)$ of the PLL is

$$A(s) = \frac{I_{CP} \cdot K_{VCO}}{2\pi \cdot sN_2} \cdot \frac{1}{sC_z} \cdot \frac{(1 + s\tau_z)}{(1 + s\tau_p) \cdot (1 + s\tau_4)}, \tag{5.4}$$

therefore, the cross-over frequency, ω_c, is

$$\omega_c \approx \frac{I_{CP} \cdot K_{VCO} \cdot B \cdot R_p}{2\pi \cdot N_2}. \tag{5.5}$$

To achieve a system phase margin of 60°, the values of the time constants are chosen as (Craninckx and Steyaert, 1998)

$$\tau_z = \frac{1}{\omega_z} = \frac{4}{\omega_c}$$
$$\tau_p = \frac{1}{\omega_p} = \frac{1}{6\omega_c}. \tag{5.6}$$

τ_4 can be chosen to be larger than τ_p to achieve more noise suppression. However, it would reduce the phase margin. As a good compromise, τ_4 is typically designed

Table 5.1 *Components parameters used in the PLL of Fig. 5.1*

Components parameters	Value
Input clock frequency	100 MHz
Output clock frequency	3.2 GHz
Forward path division ratio N_1	1
Feedback path division ratio N_2	32
I_{cp}	1 μA
Charge pump current ratio B	30
K_{VCO}	$2\pi \cdot 250$ MHz
R_p	4.3 kΩ
C_p	34 pF
C_z	27 pF
R_4	4.3 kΩ
C_4	34 pF

to be the same as τ_p. As a result, if the resistor R_4 is γ times smaller than R_p, C_4 should be approximately γ times larger than C_p.

The passive components used in the loop filter can be expressed as follows:

$$R_p = \frac{2\pi \cdot N_2}{I_{CP}BK_{VCO}} \cdot \omega_c$$

$$C_p = \frac{1}{6R_p \cdot \omega_c} = \frac{I_{CP}BK_{VCO}}{12\pi \cdot N_2\omega_c^2}$$

$$C_2 = \frac{4}{R_1 B\omega_c} = \frac{2I_{CP}BK_{VCO}}{\pi \cdot N_2\omega_c^2} \tag{5.7}$$

$$R_4 = \frac{2\pi \cdot N_2}{I_{CP}BK_{VCO}} \cdot \frac{\omega_c}{\gamma}$$

$$C_4 = \frac{\gamma I_{CP}BK_{VCO}}{12\pi \cdot N_2\omega_c^2}.$$

With the loop bandwidth set to 200 kHz, the component parameters in the loop filter can be calculated using Equations (5.2) to (5.7). The component parameters used in the synthesizer are summarized in Table 5.1.

5.3. Mathematical model

The mathematical model, as shown in Fig. 5.4, can be used to investigate the transfer function of each building block and how each contributes to the overall transfer function of the whole synthesizer system. The dividers in the forward path and feedback path have gains equal to $1/N_1$ and $1/N_2$, respectively. In addition, the gain of the PFD together with the charge pump is $I_{CP}/2\pi$. The transfer function of

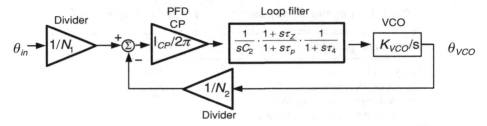

Fig. 5.4 Mathematical model for the synthesizer

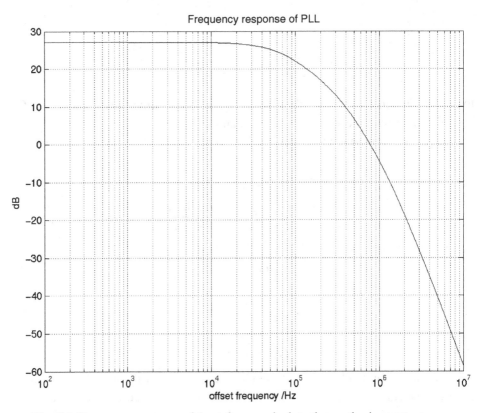

Fig. 5.5 Frequency response of the reference clock to the synthesizer output

the loop filter in the model is given by Equation (5.2). The VCO has gain equal to K_{VCO}/s.

The reference clock exhibits a low-pass characteristic as shown in Fig. 5.5, while the VCO exhibits a high-pass characteristic as shown in Fig. 5.6. Both frequency responses are determined by the loop bandwidth. The high-pass characteristic of the VCO frequency response indicates that the phase noise of the VCO within the loop bandwidth is suppressed by the PLL.

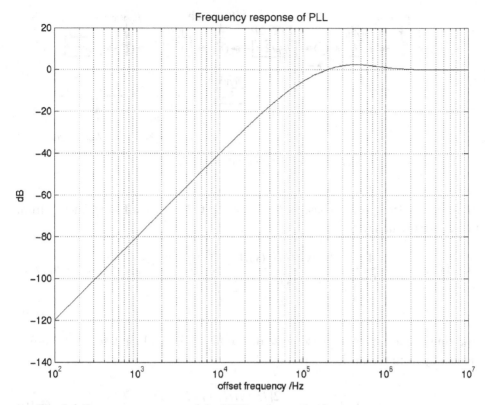

Fig. 5.6 Frequency response of the VCO to the synthesizer output

Every component in the loop generates noise which contributes to the overall phase noise at the output of the synthesizer. Noise mixes with the carrier and appears as sidebands in the frequency domain whose power mainly depends on the loop bandwidth of the synthesizer. A larger loop bandwidth could reduce the timing error of the VCO in a short period. However, this would allow more noise from the input source to appear at the output of the synthesizer. On the other hand, a smaller loop bandwidth can help reduce the input jitter and suppress the noise coming from the input clock and the charge pumps. Therefore, different building blocks in the loop have different responses to the synthesizer. Based on the linear model shown in Fig. 5.7, the total noise contribution can be obtained as follows.

Let

$$H_i(s) = \frac{1}{sC_Z},$$

$$H_L(s) = \frac{R_p}{1 + sR_pC_p}, \tag{5.8}$$

$$H_4(s) = \frac{1}{1 + sR_4C4}.$$

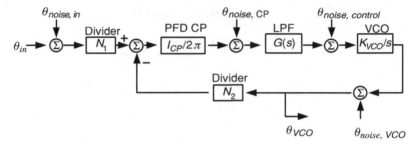

Fig. 5.7 Linear model of the synthesizer with noise source

The single-sided spectral phase noise due to the CPs at $\Delta\omega$ offset frequency from the carrier frequency is

$$L_{CP}(s)|_{s=j\Delta\omega} = \frac{4}{B^2} \frac{H_L^2(s) + H_i^2(s)B^2}{\left(s + \dfrac{I_{CP} \cdot K_{VCO} \cdot G(s)}{2\pi N_2}\right)^2} \cdot H_4^2(s) \cdot K_{VCO}^2 \cdot \lambda_{on} \cdot kT \cdot g_{m,CP1}. \tag{5.9}$$

The single-sided spectral phase noise due to the R_p in the loop filter at $\Delta\omega$ offset frequency from the carrier frequency is

$$L_{Rp}(s)|_{s=j\Delta\omega} = \frac{2[H_L(s)H_4(s) \cdot K_{VCO}]^2}{\left(s + \dfrac{I_{CP} \cdot K_{VCO} \cdot G(s)}{2\pi N_2}\right)^2} \frac{kT}{R_p}. \tag{5.10}$$

The single-sided spectral phase noise due to the R_4 in the loop filter at $\Delta\omega$ offset frequency from the carrier frequency is

$$L_{R4}(s)|_{s=j\Delta\omega} = \frac{2[H_4(s) \cdot K_{VCO}]^2 \cdot kT R_4}{\left(s + \dfrac{I_{CP} \cdot K_{VCO} \cdot G(s)}{2\pi N_2}\right)^2}. \tag{5.11}$$

The single-sided spectral phase noise due to the op amp in the loop filter at $\Delta\omega$ offset frequency from the carrier frequency is

$$L_{OP}(s)|_{s=j\Delta\omega} = \frac{1}{2} \frac{[H_4(s) \cdot K_{VCO}]^2 \cdot N_{OP}(s)}{\left(s + \dfrac{I_{CP} \cdot K_{VCO} \cdot G(s)}{2\pi N_2}\right)^2}, \tag{5.12}$$

where $N_{OP}(s)$ is the noise voltage power-spectral density of the op amp.

The single-sided spectral phase noise due to input noise at $\Delta\omega$ offset frequency from the carrier frequency is

$$L_{in}(s)|_{s=j\Delta\omega} = \frac{1}{8N_i^2\pi^2} \frac{[I_{CP}H_i(s)H_L(s)H_4(s) \cdot K_{VCO}]^2 \cdot N_{in}(s)}{\left(s + \dfrac{I_{CP} \cdot K_{VCO} \cdot G(s)}{2\pi N_2}\right)^2}, \quad (5.13)$$

where $N_{in}(s)$ is the noise power-spectral density of the input clock.

The single-sided spectral phase noise due to the noise of the VCO at $\Delta\omega$ offset frequency from the carrier frequency is

$$L_{VCO}(s)|_{s=j\Delta\omega} = \frac{1}{2} \frac{N_{VCO}(s)}{\left(s + \dfrac{I_{CP} \cdot K_{VCO} \cdot G(s)}{2\pi N_2}\right)^2}, \quad (5.14)$$

where $N_{VCO}(s)$ is the noise power-spectral density of the VCO.

After the derivation of the phase noise due to each building block in the loop, the total phase noise of the PLL can be expressed as follows:

$$L_{PLL}(s)|_{s=j\Delta\omega} = L_{VCO}(s) + L_{in}(s) + L_{OP}(s) + L_{R4}(s) + L_{RP}(s) + L_{CP}(s). \quad (5.15)$$

Figure 5.8 plots the individual noise contribution of each building block together with the total phase noise for the whole synthesizer system. From the figure, it can be observed that the CP's noise has a low-pass characteristic. The resistors in the loop filter perform a bandpass function, while the VCO exhibits a high-pass characteristic. It can be concluded that the phase noise of the VCO is attenuated within the loop bandwidth while there is no suppression at the offset frequency larger than the loop bandwidth. As a result, the VCO's in-band phase noise can be attenuated more with a larger loop bandwidth. On the other hand, a smaller loop bandwidth is able to suppress the noise contribution from the CP. In addition, the resistors and the op amp in the loop filter can be designed to have insignificant noise contribution.

5.4. Behavioral model using AC analysis

A PLL or a synthesizer is a non-linear system when operating out of its locked range. With the aid of frequency acquisition, the synthesizer can reach within the locking range where it can be assumed to be a linear system (Best, 1999). An AC analysis is a fast simulation process which can also be used to simulate the stability of PLL and synthesizer systems in the phase domain (Gutierrez, Kong and Coy, 1998). Based on the block diagram of the synthesizer shown in Fig. 5.2 and the

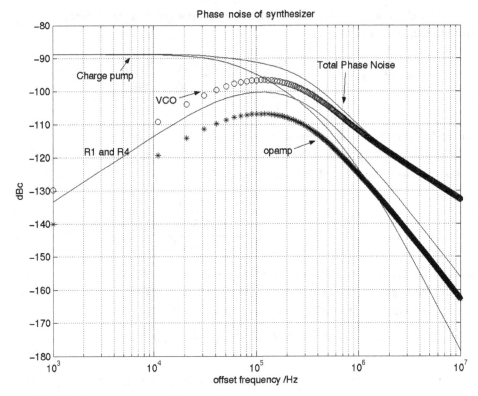

Fig. 5.8 Phase noise plot of the synthesizer

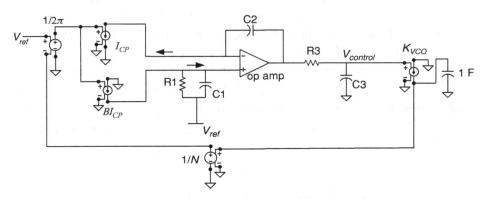

Fig. 5.9 Simplified loop stability model

component parameters given in Table 5.1, the simplified loop stability model and
the AC response are shown in Fig. 5.9 and Fig. 5.10, respectively.

The behavioral model consists of a voltage-controlled voltage source, with a
gain equal to $1/2\pi$, to model the phase frequency detector. The CP is realized by
a voltage-controlled current source which is followed by the loop filter. The loop

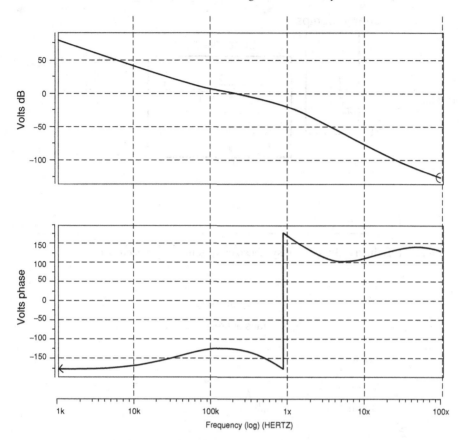

Fig. 5.10 Open-loop response of the stability model

filter is constructed using resistors and capacitors. The VCO is modeled by using a voltage-controlled current source together with a 1 F capacitor for integration. The gain of the voltage-controlled current source is equal to the VCO gain K_{VCO}. After the VCO, another voltage-controlled voltage source with the gain set to $1/N$ is used to model the divider.

At the cross-over frequency of around 200 kHz, the phase margin shown is $60°$. This ensures that the loop is stable.

5.5. Behavioral model using transient analysis

After performing AC analysis to guarantee that the loop achieves a sufficiently good phase margin for stability, a transient analysis can be applied to the behavioral model to check the loop dynamic behavior. This kind of analysis can be simulated using Spectre and SpectreRF (Kundert, 1995) which are available from Cadence Design

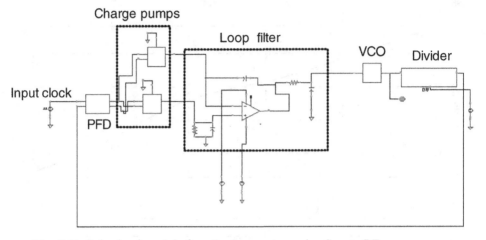

Fig. 5.11 Behavioral model of synthesizer system using SpectreRF

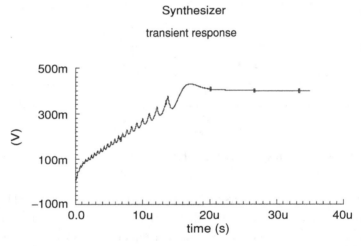

Fig. 5.12 Transient simulation result of the synthesizer's behavioral model

Systems. The schematic of the synthesizer under this simulation environment can be used for both simulations at the transistor level and system level simultaneously. Consequently, design effort in debugging and identifying which building block in the loop may cause malfunction to the system can be minimized. A block diagram of the synthesizer using ideal building blocks that are written and presented by Verilog-A in Cadence's SpectreRF is depicted in Fig. 5.11.

Figure 5.12 shows the transient response of the control voltage during locking. In frequency acquisition, cycle slipping can be observed repeatedly. Since the frequency difference of the two input clocks to the PFD is large, the PFD changes

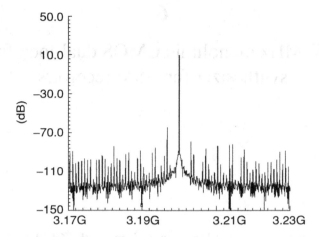

Fig. 5.13 Output spectrum of the synthesizer using the behavioral model

its output across the PFD's transfer function from 0 to $\pm 2\pi$. This causes cycle slipping. The frequency difference is continuously being reduced during frequency acquisition. When the control voltage is adjusted within the pull-in range, no cycle slipping occurs and both the frequency and phase of the VCO are eventually locked. The frequency spectrum of the synthesizer under locking is illustrated in Fig. 5.13.

6

A 2 V 900 MHz monolithic CMOS dual-loop frequency synthesizer for GSM receivers

This chapter introduces a dual-loop synthesizer for the GSM 900 specification using the 0.5 μm CMOS process with a supply voltage of 2 V. This architecture can show the advantage of providing narrow channel selection but offering low phase noise and fast switching time. The performance of the synthesizer achieves a high operating frequency (935.2–959.8 MHz), low power consumption (34 mW), low phase noise (−121.8 dBc/Hz at 600 kHz), low spurious level (−82.0 dBc at 11.3 MHz) and fast switching time (830 μs).

6.1. Design specification

The performance of frequency synthesizers is mainly specified by their output frequency, phase noise, spurious level and switching time. This section derives the specifications of a frequency synthesizer for GSM receivers.

6.1.1. Output frequency

In GSM 900 systems, the receiver-channel frequencies are expressed as follows:

$$f_{RF} = 935.2 + 0.2(N - 1)\,\text{MHz}, \tag{6.1}$$

where $N = 1, 2, \ldots, 124$ is the channel number. To receive signals in different channels, a GSM-receiver front-end shown in Fig. 6.1 is adopted. The receiver front-end consists of a low noise amplifier (LNA) and an RF filter for filtering out-of-band noise and blocking signals. The received signal is then mixed down to an IF frequency (f_{IF}) of 70 MHz for base-band signal processing. To extract information from the desired channel, the local oscillator (LO) output frequency (f_{LO}) of the frequency synthesizer is changed accordingly as follows:

$$f_{LO} = 865.2 + 0.2(N - 1) = 865.2 - 889.8\,\text{MHz}, \tag{6.2}$$

which is the output-frequency range of the frequency synthesizer to be achieved.

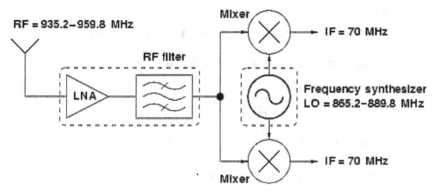

Fig. 6.1 Block diagram of the GSM receiver front-end

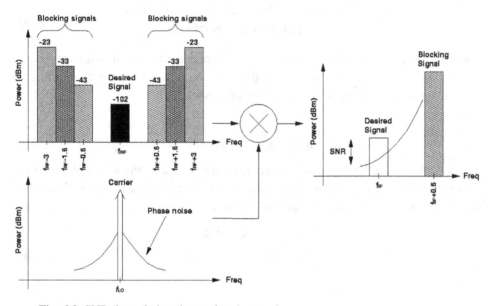

Fig. 6.2 SNR degradation due to the phase noise

6.1.2. Phase noise

The blocking-signal specification for GSM 900 receivers is shown in Fig. 6.2, where the desired-signal power can be as low as -102 dBm. At 600 kHz offset frequency, the power of the blocking signal is up to -43 dBm (ETSI, 1996). Because of the noise of MOS transistors, inductors and varactors, phase noise of typical VCOs and synthesizers appears like a skirt around the carrier signal.

To derive the phase-noise specification of the synthesizer, the SNR degradation due to the phase noise is considered. Assume that the conversion gains of the mixer and the power-spectral density of the phase noise are relatively flat, the phase-noise

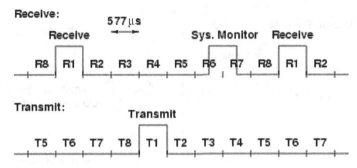

Fig. 6.3 GSM 900 receive and transmit time

specification, $L_{spec}(\Delta\omega)$, can be expressed as follows:

$$L_{spec}(\Delta\omega) < S_{dec} - S_{blk} - \text{SNR}_{spec} - 10\log(f_{ch})$$
$$= -121 \text{ dBc/Hz at } 600 \text{ kHz},$$

(6.3)

where an SNR_{spec} of 9 dB is the minimum SNR specification for the whole receiver, and $S_{des} = -102$ dBm and $S_{blk} = -43$ dBm are the power levels of the minimum desired signal and maximum blocking signal, respectively.

6.1.3. Spurious tones

The derivation of the spurious-tone specification is similar to that of the phase noise except that the channel bandwidth is not considered in this case. The spurious-tone specification, S_{spec}, can be expressed as follows:

$$S_{spec} < S_{dec} - S_{blk} - \text{SNR}_{spec}$$
$$= -68 \text{ dBc at } 1.6 \text{ MHz}$$

(6.4)

and

$$= -88 \text{ dBc for offset} > 3 \text{ MHz}.$$

6.1.4. Switching time

Although GSM 900 is globally a frequency-division-multiple-access (FDMA) system, time-division-multiple-access (TDMA) is adopted within each frequency channel. As shown in Fig. 6.3, each frequency channel is divided into eight time slots. The signal is received in time slot 1 and then transmitted in time slot 4. For system monitoring purposes, a time slot in between slot 6 and slot 7 is occupied. The most critical switching time is from the transmission period (slot 4) to the system-monitoring period (between slot 6 and slot 7). Therefore, the switching-time

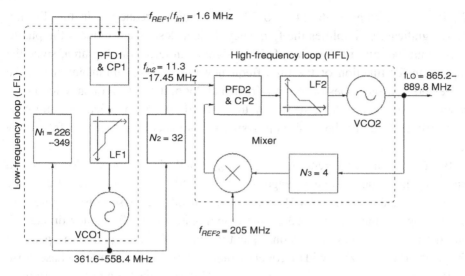

Fig. 6.4 The proposed dual-loop frequency synthesizer

requirement of the frequency synthesizer is 1.5 time slots which is equal to 865 μs.

6.1.5. Dual-loop design

To reduce the switching time and the chip area of a synthesizer, a high loop bandwidth and a high input-reference frequency are required. Moreover, to improve frequency-divider complexity, a lower frequency-division ratio is desirable. Therefore, a dual-loop frequency synthesizer is proposed (Aytur and Khoury, 1997). As shown in Fig. 6.4, the dual-loop design consists of two reference signals and two PLLs in cascade configuration (Yan and Luong, 2001). When both PLLs lock, the output frequency of the synthesizer is expressed as

$$f_{LO} = N_3 f_{REF2} + N_1 \left(\frac{N_3}{N_2} \right) f_{REF1}, \tag{6.5}$$

where f_{REF1} and f_{REF2} are the frequencies of the two reference signals, and N_1, N_2 and N_3 are frequency division ratios.

Owing to the dual-loop architecture, the input frequencies of the low-frequency and high-frequency loops are scaled up from 200 kHz to 1.6 MHz and 11.3 MHz, respectively. Therefore, the loop bandwidths of both PLLs can be increased so that the switching time and the chip area can be reduced. Compared with single-loop integer-N designs, the frequency-division ratio of the programmable divider

N_1 is reduced from $4236:4449$ to $226:349$. Such a reduction in the division ratio significantly simplifies the frequency-divider design and reduces the phase-noise contribution of the input reference signals. In terms of chip area, since the input transfer function of the high-frequency loop greatly attenuates the phase noise and spurious tones of the low-frequency loop, the low-frequency loop only requires a small loop filter to satisfy the relaxed design specification, and consequently the area of the dual-loop design is comparable to that of a fraction-N design.

Because the input-reference frequency, f_{REF1}, of the low-frequency loop is scaled up eight times, the frequency range of the oscillator VCO1 in the low-frequency loop is also scaled up to 200 MHz. On the other hand, the phase-noise requirement of the ring oscillator is relaxed by the combination of the frequency divider N_2 and the high-frequency loop. Consequently, this VCO requires a high operating frequency (600 MHz), a wide frequency range (200 MHz) and a low phase noise (-103 dBc/Hz at 600 kHz). A novel ring VCO design has been proposed to meet this tough specification (Yan and Luong, 2001).

6.2. Circuit implementation

This section discusses the circuit implementation of all the building blocks required for the proposed dual-loop synthesizer, including VCOs, frequency dividers, PFDs, charge pumps, low-pass filters and the mixer.

6.2.1. Oscillator VCO2

6.2.1.1. Architecture

Since the far-offset phase noise is dominated by VCO2, the LC oscillator needs to meet the stringent phase-noise specification. Figure 6.5 shows the schematic of the LC oscillator. Cross-coupled transistors M_{n1} are used to start and to maintain oscillation. The *pn*-junction varactors are used for frequency-tuning. The 1.1 V common-mode output voltage is designed to drive the frequency divider N_3. To reduce the phase-noise contribution due to flicker noise, PMOS transistors M_{b1} are used as a current source.

6.2.1.2. Center frequency and power consumption

Figure 6.6 shows the linear circuit model, which is used to calculate the center frequency and power consumption of the LC oscillator. The output impedance

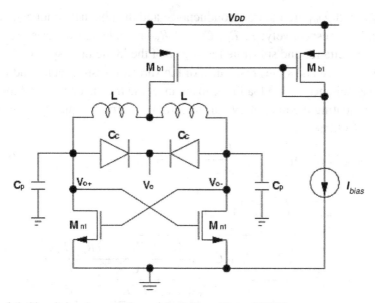

Fig. 6.5 Circuit implementation of the LC oscillator VCO2

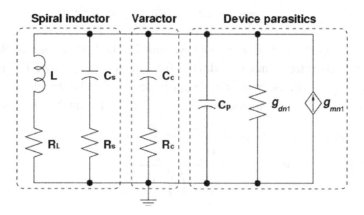

Fig. 6.6 Linear circuit model for the LC oscillator analysis

$Y_{LC}(\Delta\omega)$ of the oscillator is expressed as

$$Y_{LC}(\Delta\omega) = \left\{ \frac{R_{\mathrm{L}}}{R_{\mathrm{L}}^2 + (\omega L)^2} + \frac{R_s(\omega C_s)^2}{1 + (\omega R_s C_s)^2} + \frac{1}{R_P} + \frac{R_c(\omega C_c)^2}{1 + (\omega R_c^2 C_c^2)} \right\}$$

$$+ j\omega \cdot \left\{ \frac{L}{R_{\mathrm{L}}^2 + (\omega L)^2} + \frac{C_s}{1 + (\omega R_s C_s)^2} + C_p + \frac{C_c}{1 + (\omega R_c^2 C_c^2)} \right\},$$ (6.6)

where g_{mn1} and g_{dn1} are the transconductance and the channel conductance of the transistor M_{n1}, respectively; L, R_L, C_s and R_s are inductance, series resistance, substrate capacitance and substrate resistance of the inductor, respectively; C_c and R_c are capacitance and series resistance of the varactor, respectively; and C_p is the parasitic capacitance due to loading capacitance, the transistor M_{n1} and the output buffer. By equating the imaginary part to zero, the output frequency, f_o, of the LC oscillator VCO2 is

$$imag[Y(\Delta\omega)] \approx \frac{-L}{R_L^2 + (\omega L)^2} + C_s + C_p + \frac{C_c}{1 + (\omega R_c^2 C_c^2)} = 0$$

$$f_o = \frac{1}{2\pi\sqrt{L(C_c + C_s + C_p)}}$$

$$\cdot \sqrt{1 - \frac{(C_c + C_s + C_p) + \frac{(C_s + C_p - L/R_L^2)(R_c C_c)^2}{L(C_c + C_s + C_p)}}{\frac{L}{R_L^2} + \frac{(C_s + C_p - L/R_L^2)(R_c C_c)^2}{L(C_c + C_s + C_p)}}} \quad (6.7)$$

$$f_o = \frac{1}{2\pi\sqrt{L(C_c + C_s + C_p)}} \quad \text{when} \quad R_L = R_C = 0.$$

The first term is the center frequency without LC tank loss. Due to the series resistors, the center frequency may deviate from the ideal value by 20%. To ensure oscillation start-up, transconductance g_{mn1} is designed to be double the LC tank loss. As such, the power consumption of the LC oscillator can be expressed as

$$g_{mn1} = \sqrt{2\mu_n C_{ox}(w/L)_n I_{dn1}} = 2G_{m\_min}$$

$$\text{Power} = 2V_{DD}I_{dn1} = \frac{4V_{DD}G_{m\_min}^2}{\mu_n C_{ox}\left(\dfrac{W}{L}\right)_{n1}} \quad (6.8)$$

$$= \frac{4V_{DD}}{\mu_n C_{ox}\left(\dfrac{W}{L}\right)_{n1}} \cdot \left\{ \frac{R_L}{R_L^2 + (\omega L)^2} + \frac{R_s(\omega C_s^2)}{1 + (\omega R_s C_s)^2} \right.$$

$$\left. + \frac{1}{R_p} + \frac{R_c(\omega C_c)^2}{1 + (\omega R_c^2 C_c^2)} \right\},$$

where μ_n is the NMOS mobility constant, C_{ox} is the oxide capacitance, and W_{n1} and L_{n1} are the channel width and length of transistor M_{n1}, respectively. To minimize the power consumption, a large inductance, L, with a small series resistance R_L is preferred. Moreover, the *pn*-junction varactors with high quality factor and larger transistor M_{n1} also reduce power consumption.

6.2.1.3. Phase noise

The phase-noise estimation of the LC oscillator is based on the method by Hajimiri and Lee (1998). By lumping all the noise current into an equivalent noise current source, the phase noise of the LC oscillator, VCO2, can be expressed as

$$L_{VCO2}(\Delta\omega) = 10 \log\left(\frac{\Gamma_{rms}^2}{C_L^2 V_p^2} \cdot \frac{\overline{i_n^2}/\Delta f}{2\Delta\omega^2}\right), \tag{6.9}$$

where C_L is the total parallel capacitance, V_p is the peak voltage, Γ_{rms} is the root mean square of the impulse-stimulus function, and $\overline{i_n^2}$ is the power spectral density of the equivalent noise current. Including the noise contributed from transistors M_{n1} and M_{b1}, the spiral inductor, the *pn*-junction varactor and substrate parasitics, the total noise-power-spectral density can be expressed as

$$\frac{i_n^2}{\Delta f} = 2 \cdot 4kT \left(3 \cdot g_{mn1} + 3 \cdot \frac{g_{mb1}}{2} + \frac{1}{R_L(1 + Q_L^2)}\right.$$
$$\left. + \frac{1}{R_s(1 + Q_s^2)} + \frac{1}{R_C(1 + Q_C^2)}\right)$$

$$\tag{6.10}$$

$$Q_L = \frac{\omega L}{R_L} \quad Q_s = \frac{1}{\omega R_s C_s} \quad Q_C = \frac{1}{\omega R_C C_C},$$

where Q_L, Q_s and Q_C are the quality factors of the inductor, the inductor parasitic capacitor and the varactor, respectively.

The flow chart in Fig. 6.7 illustrates how to obtain the value Γ_{rms} in simulation. It can be done by injecting current pulses into the two output nodes of the VCO as the noise source. The total area of current impulse injected is equivalent to the total charge Δq injected into the nodes, which causes a corresponding voltage change, ΔV, as well as a time shift, Δt, at the output. The equivalent phase change $\Delta\phi$ is then calculated by

$$\Delta\phi = 2\pi \frac{\Delta t}{T}, \tag{6.11}$$

where T is the period of the oscillating waveform.

Additionally, the phase change is proportional to the injected charge:

$$\Delta\phi = \Gamma(\omega_o t) \frac{\Delta q}{q_{swing}}, \tag{6.12}$$

where q_{swing} is equal to $C_{out} V_{swing}$, V_{swing} is the voltage across the capacitor C_{out}, $\Gamma(\omega_o t)$ is the impulse sensitivity function, and the rms of $\Gamma(\omega_o t)$ is Γ_{rms}.

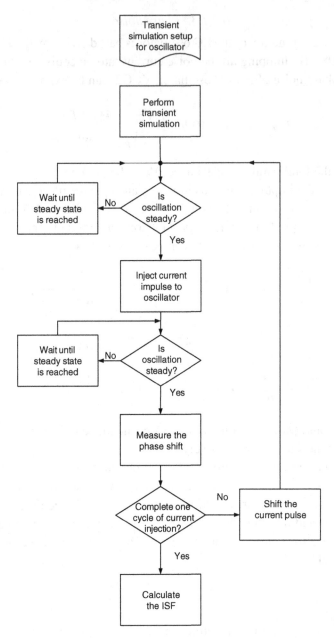

Fig. 6.7 Flowchart showing how to estimate ISF parameters

Since the impulse sensitivity function (ISF) is a periodic function with frequency ω_o, the Fourier series can be expressed as

$$\Gamma(\omega_o t) = \frac{c_0}{2} + \sum_{m=1}^{\infty} c_m(m\omega_o t + \theta_m), \tag{6.13}$$

where c_m are the coefficients of the mth harmonic and θ_m is the phase of the mth harmonic.

The c_0 term accounts for the average value of ISF, which determines the up-conversion of flicker noise. In order to minimize flicker noise, c_0 should be minimized by designing the output waveform to be symmetrical. In other words, the rising time and falling time of the output waveform should be as close as possible. On the other hand, the coefficients of c_m account for the up-conversion of white noise and are determined by the duty cycle of the waveform. The coefficients are larger if the duty cycle of the waveform is not 50% (Hajimiri and Lee, 1998). LC oscillators can achieve a 50% duty-cycle and thus have less up-conversion of white noise compared with ring oscillators.

6.2.1.4. Design issues

To design an LC oscillator that satisfies the phase-noise requirement with minimum power consumption according to Equation (6.8), inductors with large inductance and small series resistance are required. Therefore, two-layer inductors are adopted (Merrill *et al.*, 1995) for which the inductance and the quality factor can be scaled up by 4 and by 2, respectively. For the same reason, *pn*-junction varactors are inter-digitized to enhance the quality factor. Finally, g_{mn1} is designed so that it does not over compensate the LC tank too much (only twice) to reduce phase-noise contribution by transistors M_{n1}. By using SpectreRF (Cadence, 1998), the simulated phase noise is -124 dBc/Hz at 600 kHz.

6.2.2. Frequency divider N_2 and N_3

Frequency divider N_2 divides the output signal of the low-frequency loop by 32 and provides an overall 18 dB suppression of phase noise and spurious tone. The divide-by-32 divider consists of five cascaded divide-by-2 dividers. Each frequency divider is implemented by a True-Single-Phase-Clock (TSPC) logic D-type flip-flop as shown in Fig. 6.8.

As the divide-by-4 frequency divider N_3 needs to convert sinusoidal signals from the VCO2 output into square-wave signals, the first stage of the divider is implemented by pseudo-NMOS logic while the second divide-by-2 divider is implemented by the same TSPC-logic divide-by-2 circuit used for N_2. The first divider is shown in Fig. 6.9 and consists of a pseudo-NMOS amplifier and a pseudo-NMOS

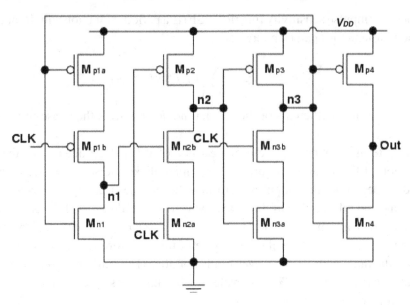

Fig. 6.8 Circuit implementation of the TSPC divide-by-2 frequency divider

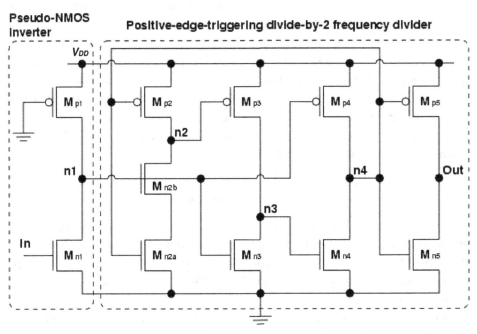

Fig. 6.9 Circuit implementation of the pseudo-NMOS divide-by-2 frequency divider

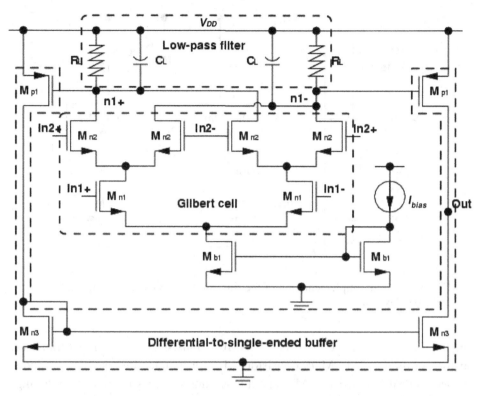

Fig. 6.10 Circuit implementation of the down-conversion mixer

divide-by-2 divider. Since the pseudo-NMOS logic is a ratioed logic, the ratio between PMOS and NMOS transistors is designed to be less than 1.6.

6.2.3. Down-conversion mixer

The mixer in the feedback path of the high-frequency loop provides the frequency shift of the synthesizer output. Figure 6.10 shows the circuit implementation of the down-conversion mixer, which consists of a Gilbert cell for mixing, a low-pass filter for eliminating high-frequency glitches, and a differential-to-single-ended buffer for amplifying the output signal into a square wave. The output bandwidth is designed to be approximately 10 MHz.

6.2.4. Ring oscillator VCO1

6.2.4.1. Architecture

Figure 6.11 shows the schematic of the proposed two-stage ring oscillator and its delay cell designed to meet the required specification as the synthesizer described in

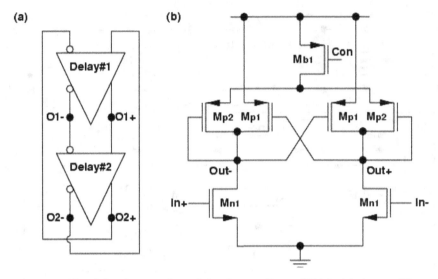

Fig. 6.11 Circuit implementation of the ring oscillator VCO1: (a) ring oscillator and (b) delay cell

Section 7.2. The delay cell consists of NMOS transistors M_{n1} as input transconductors, cross-coupled PMOS transistors M_{p1} for maintaining oscillation, diode-connected PMOS transistors M_{p2} and a bias transistor M_{b1} for frequency tuning. The source nodes of transistors M_{p1} are connected to the supply to maximize its output amplitude V_p, which also helps suppress noise sources and thus further enhance the phase-noise performance.

6.2.4.2. Output frequency

To calculate the oscillating frequency of the ring oscillator, the transfer function of the delay cell is expressed as

$$A(s) = \frac{V_o}{V_{in}} = \frac{g_{mn1}}{(-g_{mp1} + g_{mp2} + G_L) + sC_L}$$
$$G_L = g_{gn1} + g_{dp1} + g_{dp2} \tag{6.14}$$
$$C_L = C_{gsn1} + 2C_{gdn1} + C_{dbn1} + C_{gsp1} + 2C_{gdp1} + C_{dbp1}$$
$$= + C_{gsp2} + C_{dbp2} + C_{buffer},$$

where g_m is the transconductance, g_d is the channel conductance, C_{gs} is the gate-to-source capacitance, C_{gd} is the gate-to-drain capacitance, C_{db} is the drain-to-bulk capacitance, and C_{buffer} is the capacitance of the output buffer for measurement purposes. Oscillation starts when g_{mp1} is large enough to overcome the output load

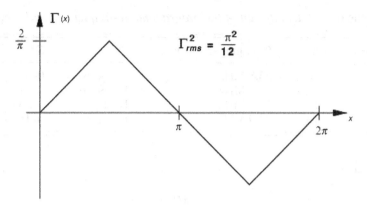

Fig. 6.12 Approximate ISF for the ring oscillator phase-noise analysis

G_L ($g_{mp1} > G_L$). By equating the delay-cell voltage gain to unity, the oscillating frequency can be expressed as

$$f_{osc} = \frac{1}{2\pi} \sqrt{\frac{g_{mn1}^2 - (-g_{mp1} + g_{mp2} + G_L)^2}{C_L^2}}. \qquad (6.15)$$

At the maximum frequency, f_{max}, the control voltage $V_{con} = 0$, and transistors M_{p2} are turned on to cancel g_{mn1}. At the minimum frequency, f_{min}, the control voltage $V_{con} = V_{DD}$, and transistors M_{p2} are turned off ($g_{mp2} = 0$). Then f_{max}, f_{min} and the frequency range, f_{range}, are expressed as

$$f_{max} \approx \frac{1}{2\pi} \cdot \frac{g_{mn1}}{C_L} \qquad f_{min} \approx \frac{1}{2\pi} \cdot \sqrt{\frac{g_{mn1}^2 - g_{mp1}^2}{C_L^2}}$$

$$f_{range} \approx f_{max} \cdot \left(1 - \sqrt{1 - \left(\frac{g_{mp1}}{g_{mn1}} \right)^2} \right). \qquad (6.16)$$

Since f_{max} is proportional to g_{mn1}/C_L, NMOS transistors are adopted to minimize power consumption. From Equation (6.16), 50% tuning range can be achieved when $g_{mp1}/g_{mn1} = 3/4$.

6.2.4.3. Phase noise

Phase-noise analysis of the ring oscillator is based on the analysis by (Hajimiri, Limotyrakis and Lee 1998). The approximate impulse-stimulus function (ISF) of the ring oscillator is shown in Fig. 6.12. The phase noise of the ring oscillator VCO1

Table 6.1 *System design of the programmable-frequency divider* N_1

Case	N	Operating frequency	P	S
1	10	55.8 MHz	22–34 (6)	0–9 (4)
2	12	46.5 MHz	18–29 (5)	0–11 (4)
3	14	39.9 MHz	16–24 (5)	0–13 (4)
4	16	34.9 MHz	14–21 (5)	0–15 (5)

is expressed as

$$\Gamma_{rms}^2 = \frac{2}{\pi} \int_0^{\pi/2} x^2 dx = \frac{\pi^2}{12}$$

$$L_{VCO2}(\Delta\omega) = N \cdot \frac{\Gamma_{rms}^2}{2 \cdot \Delta\omega^2} \cdot \frac{\overline{i_n^2}/\Delta f}{C_L^2 V_p^2},$$

(6.17)

where Γ_{rms} is the root mean square of ISF, $N = 4$ is the number of noise sources, V_p is the peak output amplitude, and $\overline{i_n^2}$ is the total device noise-power-spectral density. To enhance the phase-noise performance, the source nodes of transistors M_{p1} are connected to the supply in order to maximize output amplitude. Calculation shows that the phase noise is approximately -107 dBc/Hz at 600 kHz offset.

6.2.5. Programmable frequency divider N_1

6.2.5.1. Architecture and system design

The programmable frequency divider N_1 uses a pulse swallow frequency divider as discussed in Section 2.5.2. As the frequency-division ratio (226 : 349) can be achieved with different combinations of N, S and P, the optimal combination in terms of performance needs to be identified and chosen for the design. To implement the programmable function, the division ratio of the P-counter should be larger than that of the S-counter. To optimize the power consumption, the operating frequencies and number of bits of both the P- and S-counters should be minimized.

Table 6.1 shows the different combinations of N, P and S that can implement the desired division ratio. Case 1 requires the highest operating frequencies and number of bits for the P- and S-counters, so it is not adopted. Case 4 has the problem that the S-value is larger than the P-value. It seems that Case 3 is the best, but Case 2 is chosen as the final design because it is much easier to implement an asynchronous divide-by-12 frequency divider than a divide-by-14 divider.

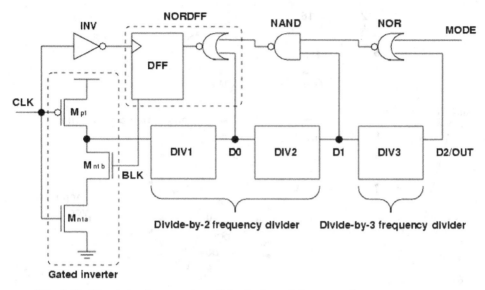

Fig. 6.13 Circuit implementation of the dual-modulus prescaler

6.2.5.2. Dual-modulus prescaler

The dual-modulus prescaler is implemented by the back-carrier-propagation approach as shown in Fig. 6.13 (Larsson, 1996). When control signal $MODE = 1$, the gated inverter is bypassed, and it is a divide-by-12 divider. When control signal $MODE = 0$, the final state $D0,1,2 = 010$ will be detected, at which $BLK = 0$ and the input signal is delayed by one-clock cycle. Thus the function of divide-by-13 is achieved. The back-carrier-propagation approach allows low-frequency signals to switch to the final state much earlier than high-frequency signals and thus reduces power consumption for a given speed.

6.2.6. Charge pumps and loop filter

Figure 6.14 shows the circuit implementation of the charge pump (CP) used in the two loops. It consists of two cascode-current sources for both the pull-up and pull-down currents, four complementary switches and a unity-gain amplifier. By using high-swing-cascode-current sources, the output impedance is increased for effective current injection. Minimum-size complementary switches are adopted to minimize clock feed-through and charge injection of the switches. The unity-gain amplifier keeps the voltages of nodes VCO and nb equal so that charge sharing between nodes VCO, ns, and ps can be minimized.

The loop filters in the two PLLs are second-order low-pass filters, which are implemented using linear capacitors and silicide-blocked polysilicon resistors. The

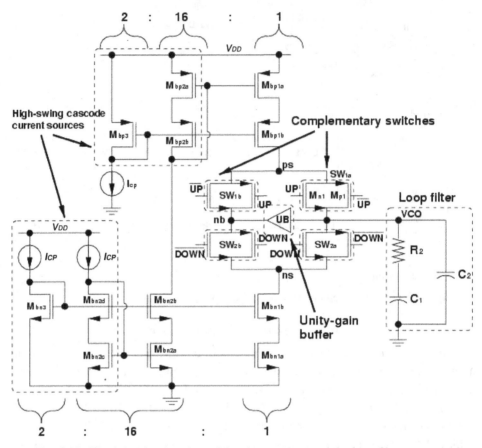

Fig. 6.14 Circuit implementation of the charge pump and the loop filter

values of capacitance, resistance and CP current are optimally designed to satisfy simultaneously the phase-noise, spurious-tone and switching-time requirements with minimum chip area. The loop bandwidth of the low-frequency and high-frequency loops are 40 kHz and 27 kHz, respectively. The phase noise of the whole synthesizer is −123.8 dBc/Hz at 600 kHz as shown in Fig. 6.15, which shows that the close-in (<100 kHz) phase noise is dominated by the charge pump CP1 and loop filter LF1, while the far-offset (>100 kHz) phase noise is dominated by the LC oscillator.

6.3. Experimental results

The dual-loop frequency synthesizer is implemented in 0.5 μm CMOS technology. Linear capacitors are put under all the bias pins to serve as on-chip bypass capacitors. Figure 6.16 shows the die photo of the dual-loop frequency synthesizer. The active area of the synthesizer is 2.64 mm^2.

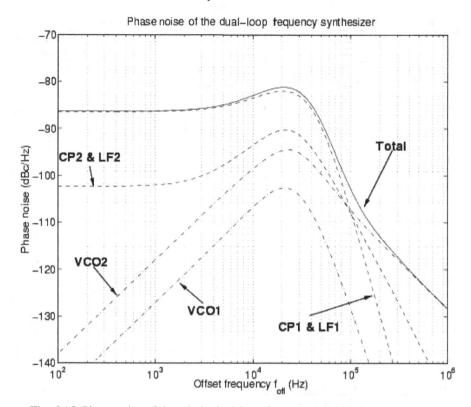

Fig. 6.15 Phase noise of the whole dual-loop frequency synthesizer

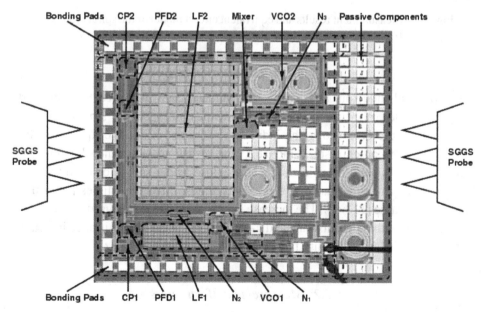

Fig. 6.16 Die photo of the dual-loop frequency synthesizer

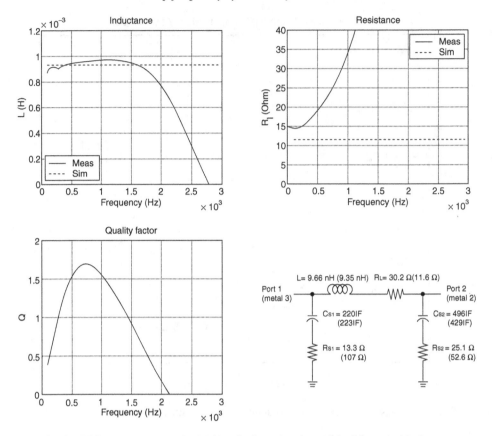

Fig. 6.17 Measurement results and equivalent circuit model of the spiral inductors at 900 MHz

6.3.1. Measurement of inductors

On-chip spiral inductors and *pn*-junction varactors are measured by a network analyzer. To de-embed the probing-pad parasitics, an open-pad structure is also measured. Figure 6.17 shows the inductance L, series resistance R_L and quality factor Q_L of the on-chip spiral inductor. The measured inductance L is close to simulation results and drops at frequencies close to the self-resonant frequency. However, the series resistance $R_L(30.2\,\Omega)$ is almost three three times larger than the expected value (11.6 Ω). The increase in series resistance is caused by eddy currents induced within the substrate and *n*-well fingers. As series resistance increases significantly, the port-1 quality factor is limited to 1.6 at 900 MHz.

6.3.2. Measurement of varactors

The *pn*-junction varactors are also measured by a network analyzer and the measurement results at 900 MHz are shown in Fig. 6.18. As the varactors are directly

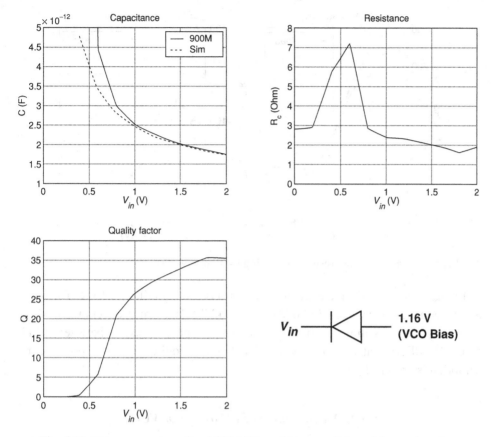

Fig. 6.18 Measurement results at 900 MHz and bias condition of the *pn*-junction varactors

connected to the output of the LC oscillator, they are biased at 1.16 V, which is the DC bias of the oscillator core. The measured capacitance, C, is close to the estimated result in the reverse-biased region. The series resistance, R_C is around 2Ω due to the minimum junction spacing and the non-minimum junction width. The quality factor is around 30 in the operating region of the oscillator.

6.3.3. Measurement of ring oscillator VCO1

The phase noise of the oscillators are measured by a direct-phase-noise measurement (Hajimiri, Limotyrakis and Lee, 1998). First, the carrier power is determined at large video (VBW) and resolution bandwidths (RBW). Then, the resolution bandwidth is reduced until the noise edges, and not the envelope of the resolution filter, are displayed. Finally, the phase noise is measured at the corresponding frequency offset from the carrier. To make sure that the measured phase noise is valid, the displayed values must be at least 10 dB above the intrinsic noise of the analyzer.

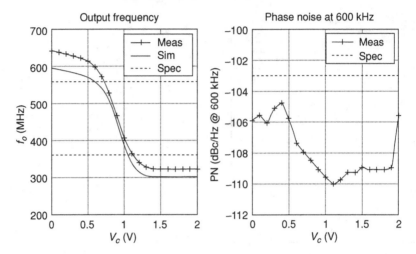

Fig. 6.19 Measurement results of the ring oscillator VCO1

Figure 6.19 shows the measurement results of the ring oscillator VCO1. The operating frequency is between 324.0 MHz and 642.2 MHz, which covers the desired frequency range. Within the frequency range, the phase noise is between -111 and -108 dBc/Hz at 600 kHz, which satisfies its requirement. The power consumption is around 10 mW.

6.3.4. Measurement of LC oscillator VCO2

Figure 6.20 shows the measurement results of the LC oscillator VCO2. Due to the quality-factor degradation of the spiral inductor, the bias current of the oscillator is increased by 15% above its designed value to achieve the phase noise specification (-121 dBc/Hz at 600 kHz). The measured operating frequency range is between 725.0 MHz and 940.5 MHz. The oscillation stops when the VCO control voltage is below 0.6 V because the varactors become forward-biased. Over the desired frequency range of between 865.2 MHz and 889.8 MHz, the achieved phase noise is below -121 dBc/Hz at 600 kHz.

6.3.5. Measurement of loop filter

Figure 6.21 shows the magnitude and phase plots of the loop-filter impedance of LF1 and LF2. When compared with the simulation results, the magnitude agrees quite well, and the phase derivation is less than $5°$. Therefore, the loop stability and the transfer functions for the phase noise and the spurious tones are well preserved. The glitches close to 1 MHz are caused by the defects in the measurement set-up since it is too close to the measurement-frequency limit.

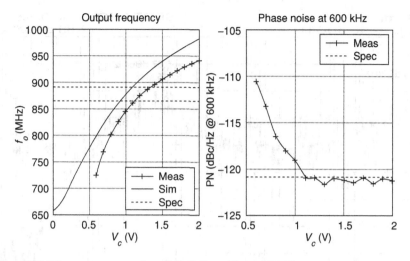

Fig. 6.20 Measurement results of the LC oscillator VCO2

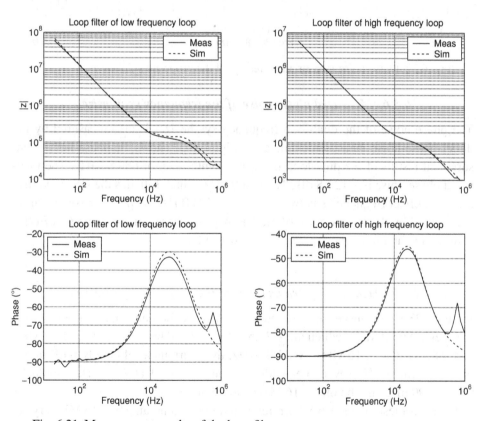

Fig. 6.21 Measurement results of the loop filters

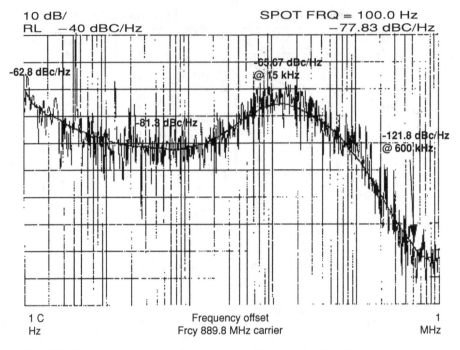

10 dB/   SPOT FRQ = 100.0 Hz
RL   −40 dBC/Hz   −77.83 dBC/Hz

-62.8 dBc/Hz

−65.67 dBc/Hz
@ 15 kHz

−81.3 dBc/Hz

-121.8 dBc/Hz
@ 600 kHz

1 C   Frequency offset   1
Hz   Frcy 889.8 MHz carrier   MHz

Fig. 6.22 Phase-noise measurement results of the proposed synthesizer

6.3.6. Measured phase noise of the frequency synthesizer

The phase noise of the dual-loop frequency synthesizer is also measured by the direct-phase-noise-measurement method. Figure 6.22 shows the phase-noise measurement results of the dual-loop frequency synthesizer at 889.8 MHz. The measured phase noise is −121.8 dBc/Hz at 600 kHz, which satisfies the GSM requirement. At offset frequencies between 10 Hz and 100 Hz, the phase noise is mainly contributed by the flicker noise of the CP. A peak phase noise of −65.67 dBc/Hz is measured at a frequency offset of around 15 kHz.

6.3.7. Measured spurious tones of the frequency synthesizer

Figure 6.23 shows the measured spurious levels of the dual-loop frequency synthesizer at 865.2 MHz, which are −79.5 dBc at 1.6 MHz, −82.0 dBc at 11.3 MHz and −82.83 dBc at 16 MHz. At 11.3 MHz, the spurious level is only 6 dB above the requirement. However, the predicted spurious level at 1.6 MHz should be below −90 dBc and the one at 16 MHz should not exist.

In fact, for testing, the 1.6 MHz reference signal is generated by a 16 MHz crystal oscillator and a decade counter. Therefore, the 16 MHz spur is caused by substrate coupling between the crystal oscillator and the synthesizer. To verify this, when the

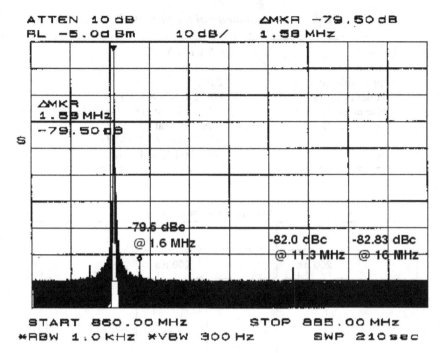

Fig. 6.23 Measured spurious level of the proposed synthesizer

low-frequency loop is disabled, the spurious level remains −75.1 dBc at 1.6 MHz, which implies that the increase in spurious level at 1.6 MHz is mainly caused by the substrate coupling.

6.3.8. Switching time of the frequency synthesizer

To measure the switching time of the frequency synthesizer, two 4-bit multiplexers are used to switch the frequency division ratio N_1 between 226 and 349. By observing the control voltage of the LC oscillator, the worst-case switching time is obtained. Figure 6.24 shows the change of the control voltages of both VCO1 and VCO2 when division ratio N_1 is switched from 226 and 349. Since the CP current of the high-frequency loop is small (0.4 μA) and the loop-filter capacitor is large (1.3 nF), the VCO control voltage is slew limited. The worst-case switching time is 830 μs, which satisfies the GSM requirement.

6.3.9. Performance evaluation

Table 6.2 summarizes the measured performance of the proposed frequency synthesizer, and Table 6.3 lists the performance of other fully-integrated synthesizers

Table 6.2 *Performance summary of the proposed synthesizer*

Process	0.5-μm CMOS
Chip area	2.64 mm^2
Supply voltage	2.0 V
Frequency range	865.2 to 889.8 MHz
Phase noise	−121.83 dBc/Hz at 600 kHz
Spurious level	−79.5 dBc at 1.6 MHz
	−82.0 dBc at 11.3 MHz
	−82.88 dBc at 16 MHz

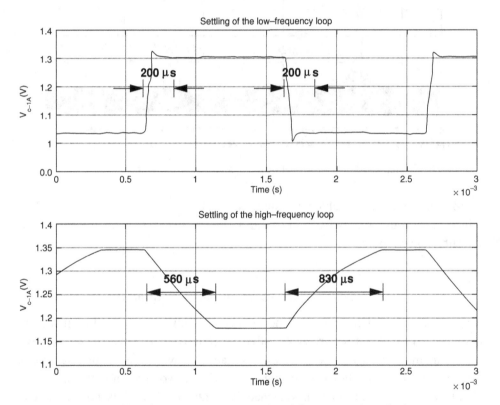

Fig. 6.24 Switching-time measurement results of the proposed synthesizer

for comparison. The proposed synthesizer operates at a single 2 V supply with a power consumption of 34 mW while all other designs require supply voltages of at least 2.7 V and consume at least 50 mW. As the fractional-N topology is adopted in the other designs, their reference frequencies are scaled up by at least 48 times. Although the reference frequency of this work is limited by the tuning range of the ring oscillator, the spurious level at 1.6 MHz still satisfies the requirement because of

Table 6.3 *Performance comparison of recent work on frequency synthesizers*

Design	(Craninckx and Steyaert, 1998)	(Ali and Tham, 1996)	(Parker and Ray, 1998)	This work
Architecture	Fractional-N	Fractional-N	Fractional-N	Dual-Loop
Process	0.4 μm CMOS	25 GHz BJT	0.6 μm CMOS	0.5 μm CMOS
Carrier frequency	1.8 GHz	900 MHz	1.6 GHz	900 MHz
Channel spacing	200 kHz	600 kHz	600 kHz	200 kHz
Reference frequency	26.6 MHz	9.6 MHz	61.5 MHz	1.6 and 205 MHz
Loop bandwidth	45 kHz	4 kHz	200 kHz	40 and 27 kHz
Chip area	3.23 mm^2	5.5 mm^2	1.6 mm^2	2.64 mm^2
600 kHz phase noise	−121 dBc/Hz	−116.6 dBc/Hz	−115 dBc/Hz	−121.83 dBc/Hz
Spurious level	−75 dBc	<−110 dBc	−83 dBc	−79.5 dBc
Switching time	<250 μs	<600 μs	N. A.	<830 μs
Supply voltage	3 V	2.7 to 5 V	3 V	2 V
Power	51 mW	50 mW	90 mW	34 mW

the filtering function of the high-frequency loop. The proposed synthesizer consists of two loop filters, but the chip area is just a little bit larger than that of Craninckx and Steyaert's design (1998) due to the use of linear capacitors. Compared with the Craninckx and Steyaert's and Parker and Ray's designs (1998), the spurious levels are between −75 dBc and −85 dBc, which indicates that both designs may suffer from the same problem with substrate coupling from the reference signal.

7

A 1.5 V 900 MHz monolithic CMOS fast-switching frequency synthesizer for wireless applications

This chapter introduces the fractional-N synthesizer, which is also aimed at GSM applications with a supply voltage of 1.5 V. The synthesizer employs a switchable-capacitor array to tune the output frequency and a dual-path loop filter operating in the capacitance domain is proposed. It provides many advantages, including simplified analog circuitry, a low-supply voltage, low power consumption, small chip area, fast frequency switching, and high immunity of substrate noise. Implemented in a standard 0.5 μm CMOS process, a fully integrated fractional synthesizer prototype with a third-order sigma–delta modulator is designed for 1.5 V and consumes 30 mW. The total chip area is around 1.0 mm^2. The settling time is less than 250 μs, and the phase noise is better than -115 dBC/Hz at 600 kH$_3$ offset.

7.1. Introduction

In monolithic phase-locked-loop (PLL) frequency synthesizer design, the phase noise performance of the synthesizer is degraded not only by the phase noise of the voltage-controlled oscillator (VCO) itself and the noise from the loop filter, but also by the substrate noise. Since the substrate is conductive, any noise generated from other circuits will couple through the substrate to the VCO and degrade the phase noise. This noise source is difficult to predict and cannot be prevented or reduced significantly even by increasing power consumption. Large separation from noise sources and guard rings can help reduce substrate coupling, but the effectiveness is quite limited in practice due to the compact layout for a small chip area.

Switching speed is limited in PLL-based synthesizers. In conventional integer-N designs, the highest frequency resolution is limited by the reference frequency. However, in order to fulfill the requirement of stability, the loop bandwidth is limited to less than approximately $1/10$ of the reference frequency (Lee, 1998). As a result, PLL synthesizers with a fine frequency resolution have a small loop bandwidth and thus a low switching speed. Moreover, at different output frequencies, the varactor of the VCO is biased with different voltages and has different gains. The loop gain is therefore not constant throughout the whole output frequency range. In some frequency ranges, the loop bandwidth and the speed are smaller than the optimum because the stability of the loop has to be ensured in all cases. Although some linearization techniques, for example using piecewise-linear gain compensation can be used (Craninckx and Steyaert, 1998), they require the characteristic of a pre-measured varactor and thus can provide only limited linearization.

To compensate for the center frequency variation in voltage-controlled oscillators, a large tuning range is required. In this case, a large voltage range is needed to bias the varactor. Thus, a high supply voltage is needed. To meet the tough phase noise specification, huge capacitors are needed to reduce the thermal noise in the CP and loop filter. The capacitors are usually so large that they will either occupy a very large chip area or have to be put off-chip.

In this work, we propose a new architecture to solve the above problems, such as substrate noise, frequency switching speed, supply voltage and chip area, which are faced by existing monolithic frequency synthesizers.

7.2. Proposed synthesizer architecture

One approach to increasing the switching speed of a PLL synthesizer is to predict the settled tuning voltage of the VCO. By adding the predicted voltage offset from a digital-to-analog converter (DAC) to the CP or loop filter output, as shown in Fig. 3.7(a), the capacitance of the VCO's LC tank and the output frequency change immediately because the loop does not need to change and settle down (Goldberg, 1996, pp. 183–4). Although the capacitance-to-frequency relationship, which is governed by the equation $f = 1/\sqrt{LC}$, is quite linear within a small range, the voltage-to-capacitance relationship of the varactor, which depends on doping and geometry, is non-linear and difficult to predict. Therefore, it is difficult to generate the required voltage to tune the VCO to the correct frequency. Consequently, only coarse tuning can be provided by the DAC. The loop still has to change a lot and settle down before the desired output frequencies are reached. Moreover, in order to reduce the noise from the DAC and the voltage adder reaching the VCO, the voltage summation has to be placed before the loop filter instead of directly before the VCO. As a result, the predicted tuning voltage still has to pass through the loop filter and take time to reach the VCO.

Based on the above approach, a new architecture is proposed in this work. Instead of doing the addition in the voltage domain, the addition is done in the capacitance domain (Lo and Luong, 2002). The addition is easily implemented by putting two capacitors in parallel. Originally, the voltages from the DAC and CP are added by a voltage adder, and the resultant voltage controls the capacitance of the LC tank of the

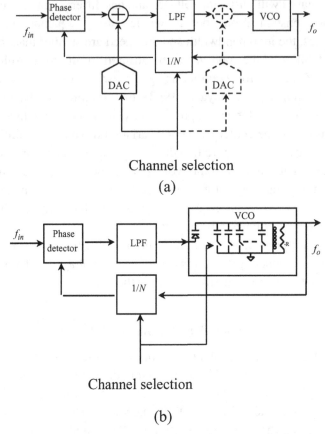

Channel selection

(a)

Channel selection

(b)

Fig. 7.1 Fast-switching PLL frequency synthesizer: (a) existing solution employing DAC to increase frequency switching speed; (b) the proposed solution employing SCA to increase frequency switching speed

VCO through a varactor with an unknown and non-linear characteristic (Fig. 7.1(a)). In the proposed architecture, the DAC is replaced by a binary-weighted switchable-capacitor array (SCA), and the voltage adder is replaced by a parallel connection of the SCA and the varactor, as shown in Fig. 7.1(b). In this case, the capacitance of the varactor controlled by the loop and the capacitance of the SCA are added. Instead of using the DAC to control the capacitance of the varactor with a non-linear voltage-to-capacitance characteristic, the capacitance of the SCA is controlled directly and linearly by the switches in the SCA. As a result, the output frequency can be controlled quite linearly, and fine tuning from the SCA is now possible. The tuning is also much faster because the control of the capacitance bypasses the loop filter. Although the SCA is also reported in other designs (Behbahani and Abidi, 1998), it is only used for the compensation of the oscillator's frequency range shift due to the process variation, not for tuning the center frequency during the operation.

Since the tuning of frequency is mainly done by the SCA, the gain of the varactor and the change of the tuning voltage can be very small. These two features provide many other advantages. Not only is faster frequency switching achieved due to the direct control inside the VCO, but the relatively constant tuning voltage also results in a constant gain of the varactor and a constant optimal loop bandwidth and speed without any linearization technique being used. It can also prevent forward biasing of the *pn*-junction varactor used. The small-variant tuning voltage also reduces the minimal supply voltage and simplifies the circuit design of the analog signal path due to the small dynamic range required.

Moreover, more noise from the loop filter can be allowed in the proposed architecture because any change in the tuning voltage has very little effect on the capacitance of the varactor and the frequency of the VCO. As a result, much smaller on-chip loop filter capacitors can be used to filter out the noise. Since the resistance of NMOS switches in the SCA changes very little with their control voltages when they are fully turned on or off, the substrate noise can only change the total capacitance and the output frequency of the LC VCO very little through the small varactor and the SCA. This reduces degradation of the phase noise of the VCO due to unpreventable substrate noise. In addition to smaller power consumption, the absence of a noisy DAC and a voltage adder can reduce tuning voltage noise which would modulate the VCO and degrade the phase noise.

In addition, we also propose a novel idea to implement a dual-path loop filter in the capacitance domain so that the total capacitor can be further reduced by about one half.

7.3. System specification and consideration

The frequency synthesizer is designed for a GSM receiver. The phase-noise requirements are -103 dBc/Hz at 400 kHz offset and -121 dBc/Hz at 600 kHz. The frequency range is 865–890 MHz for a 70 MHz IF with 200 kHz resolution. The loop bandwidth is 80 kHz. The synthesizer is based on a third-order sigma-delta fractional-N design with a 25.6 MHz reference frequency. The whole synthesizer is operated with a single 1.5 V supply.

One drawback of the fractional-N synthesizer is that quantization noise is added to the loop because there are only a finite number of division ratios (quantization levels) used in the frequency divider to represent all the division ratios in between. If a constant fractional division ratio is required, the quantization noise generated is a single frequency tone, in which the frequency depends on the fractional division ratio. On the other hand, if a randomly-changing division ratio is required, the quantization noise generated has a white noise spectrum. To solve the problem, a sigma–delta modulator can be used to control the divider. For a fractional input,

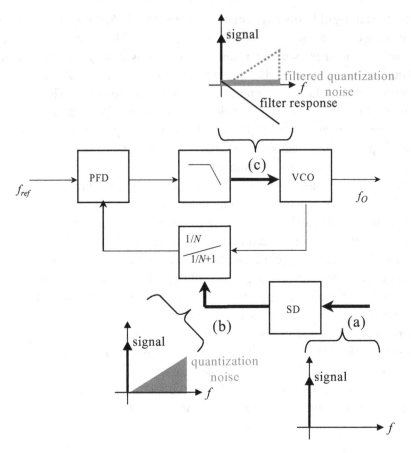

Fig. 7.2 Quantization noise in a sigma–delta fractional-N synthesizer

Fig. 7.2(a), the sigma–delta modulator can redistribute the quantization noise such that most of the noise is located at higher frequencies, Fig. 7.2(b). The resultant high frequency noise can be filtered out by the low-pass response of the loop, Fig. 7.2(c), and very little noise is left.

To design a sigma–delta fractional-N PLL synthesizer, three critical parameters, namely the reference frequency, the loop bandwidth, and the order of the sigma–delta modulator, need to be specified. In the main, these parameters determine the switching speed, the spurs and the noise of the loop.

Either a higher clock frequency, which is the same as the reference frequency of the PLL, or a higher order of the sigma–delta modulator can push the quantization noise to higher frequencies more effectively. On the other hand, a lower loop bandwidth can filter out more noise at high frequencies. To achieve a phase noise requirement of −121 dBc/Hz at 600 kHz offset, the required maximal loop bandwidths and minimal reference frequencies for the synthesizer are shown in Fig. 7.3 for second-, third- and fourth-order sigma–delta modulators. In this design,

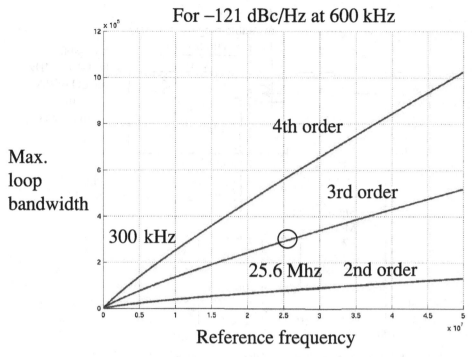

Fig. 7.3 Maximum loop bandwidths and minimum reference frequencies for second-, third- and fourth-order sigma–delta modulators for phase noise requirement of −121 dBc/Hz at 600 kHz offset

a reference frequency of 25.6 MHz and a third-order sigma–delta modulator can provide a maximal 300 kHz loop bandwidth for the required phase noise performance. The reference frequency of 25.6 MHz is used because it is 2^7 times the channel spacing, which is 200 kHz for GSM, and thus it can greatly simplify the design of the modulator. Moreover, since the reference frequency is 25.6 MHz, any spur in the output of the frequency synthesizer will be a multiple of 25.6 MHz and will be located outside the receiver band. Only the out-of-band noise, which is already greatly attenuated by the RF image rejection filter, can be mixed down to the IF by the spurs at 25.6 MHz offset.

As shown in Fig. 7.4, the complete synthesizer system includes a fractional-N PLL synthesizer, a sigma–delta modulator, and a gain-and-offset adjustment circuit. The PLL synthesizer itself consists of two quadrative LC VCOs, a novel dual-path loop filter, a CP, a PFD, a multi-modulus prescalar, and a third-order digital sigma–delta modulator.

Since a 25.6 MHz reference frequency and a prescalar with division ratios from 32 to 39.5 are used, the output frequency range is from 25.6 MHz × 32 = 819.2 MHz to 25.6 MHz × 39.5 = 1011.2 MHz, and the minimal frequency resolution is 25.6 MHz × 0.5 = 12.8 MHz. The 16 division ratios are controlled by a 4-bit digital

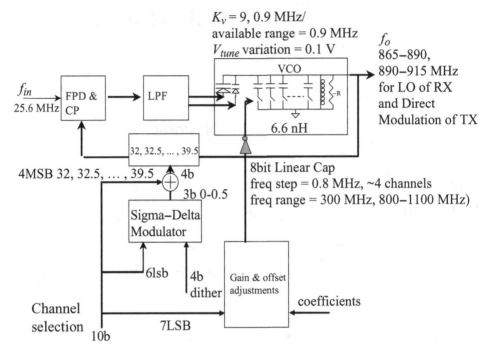

Fig. 7.4 Detailed system diagram of the proposed synthesizer

signal. In order to generate all the desired channels with a 200 kHz spacing, a digital sigma–delta modulator of at least six bits is needed because $12.8 \text{ MHz}/2^6 = 200$ kHz. Each least significant bit (LSB) of the 6-bit sigma–delta modulator represents $0.5/2^6$ in the division ratio of the prescalar. Including the four bits to directly control the prescalar, a 10-bit channel-selection word is required to control the output frequency. The extra four bits are included to add a dither signal to randomize the input of the sigma–delta modulator in order to prevent the problem of pattern noise at the output.

In a higher-order digital modulator, the output signal swing is larger than the available output value. For a third-order modulator with the output value between 0 and 0.5, the signal swing can be up to -1 and 2. Owing to the extra signal swing required, the available division ratios will be reduced. As a result, extra division ratios are included. The minimum and maximum division ratios with this prescalar will become $32 - (-1) = 33$ and $39.5 - 2 + 0.5 = 38$ and correspond to the output frequencies of 844.8 MHz and 972.8 MHz, respectively. The available output frequency range can still cover the required frequency range for both GSM receiver with IF as 70 MHz (865–890 MHz) and transmitters with direct digital modulation (890–915 MHz).

The lower seven bits of the channel-selection word represent the 125 channels required. These seven bits are used to control the SCA used in the VCOs to obtain the correct output frequency. Since these seven bits only represent the channel number but not the exact frequency of the channel, a frequency offset is added to obtain the correct frequency for the channels. Moreover, the frequency change due to a change in the least significant bit in the SCA does not exactly represent a channel. A gain adjustment is required to relate the channel number and the corresponding number of the switchable capacitors. The gain and offset adjustment circuits are implemented with digital multipliers and adders only. They can be also implemented as a few extra lines of codes in the DSP chip. Both the coefficients of the gain and the offset adjustments, which depend on the frequency step size and the center frequency of the oscillator, are affected by the process variation, and thus a calibration is needed. The self-calibration of the gain and offset adjustments can be done automatically by monitoring the tuning voltage of the VCO using a simple voltage comparator. The two coefficients are determined such that the tuning voltages of the varactor are always around a constant value throughout the whole tuning range. It can be done once at the beginning or occasionally during operation by a simple finite state machine or a few lines of DSP code.

An 8-bit SCA is used to provide a tuning range of around 300 MHz (from 800 MHz to 1100 MHz) with a frequency step of around 800 kHz. The varactors in the VCO can provide about 9 MHz/V. As a result, the tuning voltage of the varactors will vary within 0.1 V.

7.4. Circuit implementation

7.4.1. LC VCOs

The LC VCO is controlled by both the loop and the SCAs and, as shown in Fig. 7.5, includes two identical differential LC oscillators, which are mutually coupled by four coupling transistors to provide quadrature phase outputs (Rofougaran *et al.*, 1996). By sharing the current source, both the amplitude and the phase matchings of the outputs of the two oscillators are well maintained (Lo and Luong, 1999). The LC oscillator employs two-metal-layer spiral inductors (Metal 2 and Metal 3) for better quality factor and smaller area. Each oscillator consists of an 8-bit SCA as the coarse frequency tuning and two small varactors for fine frequency tuning through the loop. The SCAs employ linear capacitors and donut transistors as the switches to minimize the drain parasitic capacitance and, thus, to maximize the tuning range (Behbahani and Abidi, 1998). The sizes of the capacitors and the MOS switches in the SCA are designed to achieve a quality factor of around 15 for the SCA. The SCAs provide a tuning range of around 300 MHz and a

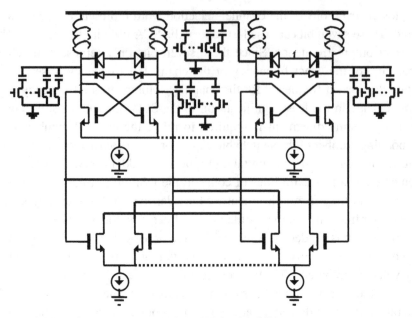

Fig. 7.5 The schematic diagram of the VCO with SCA

frequency resolution of around 800 kHz. The two small varactors have to provide the remaining 1 MHz tuning. They are designed to have small gains of 9 MHz/V and 0.9 MHz/V, and a quality factor of 25, and their tuning voltages need to vary only by 0.1 V. It is based on a parasitic *pn*-junction diode of p_+ active and *n*-well without the problem of being forward-biased because of the small tuning voltage range.

7.4.2. Loop filter

In a Type-2 PLL, a zero in the open-loop transfer function is required to maintain the loop stability. It can be implemented by a resistor in series with a capacitor in the feedback path of an active filter. However, as the capacitor, C_p, for the pole will need to be large enough to meet the noise requirement, the capacitor for the zero, C_z, will have to be much larger due to the large ratio required for the pole and the zero. As a consequence, the capacitors will be as large as a few nanofarads, which is too large to be implemented on-chip. Another way to implement the required zero with smaller capacitors is to use a dual-path filter (Craninckx and Steyaert, 1998). By adding the outputs of an integrator and of a low-pass filter, as shown in Fig. 7.6(a), the zero is formed. In this case, there is no dependency between C_z and C_p, and both of them can be minimized independently for the required noise contribution.

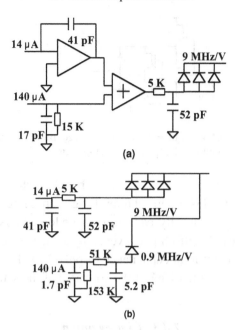

Fig. 7.6 The evolution of dual-path filter: (a) adding the outputs of the integrator and LPF with a voltage adder; (b) adding the outputs in the capacitance domain

In a similar manner to the case of adding the predicted offset from the SCA, this addition is made in the capacitance domain. Instead of using a voltage adder, which consumes power and generates noise, the two voltages are added by directly controlling two weighted varactors in the oscillator, as shown in Fig. 7.6(b). The summation is implemented in the capacitance domain as two capacitors are in parallel. Moreover, when the loop is locked, the output of the integrator is constant, and no net current flows from the charge pumps. The DC voltage generated by the low-pass filter is always zero, and so cannot contribute to the controlling and tuning of the varactor that is connected to its output. With this observation, an even smaller varactor at the output of the low-pass filter is used. In this case, the varactors used for the outputs of the integrator and of the low-pass filter have a gain of 9 MHz/V and 0.9 MHz/V, respectively.

The voltage noise from the filter can modulate the VCO to generate phase noise. However, since most of the frequency tuning is done by the SCA, the required gain and tuning range of the varactors are very small. Owing to the small varactors in the oscillator, less noise from the filter and charge pump is converted into phase noise of the VCO. Thus, larger resistance and smaller capacitors can be used in the filter.

Moreover, in this proposed filter, the path of the integrator, which can provide frequency tuning, and the path of the low-pass filter are separate. An even smaller varactor can be used with the low-pass filter path. As a result of using a smaller

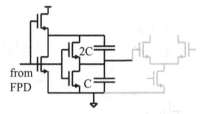

Fig. 7.7 Schematic of the current-steering charge pump (pump-down current branch only) and the switched-capacitor driving stage

varactor, more noise can be tolerated in this path and much smaller capacitors can be used there. Therefore, this proposed filter design can further reduce the required capacitors by half at most. Usually, capacitors in the loop filters occupy most of the chip area or even have to be off-chip. However, by using the SCA and the proposed dual-path filter, a resultant total capacitance of only 100 pF is needed in the loop filter, which occupies less than 10% of the core area of the synthesizer.

7.4.3. *Charge pump*

In order to provide the two paths in the loop filter with different gains, two charge pumps (CPs) of different sizes are used. Each CP is a simple current-steering type with CMOS switches operated in the saturation region for large output impedance. The output voltage range of the CPs can be very small because of the nearly constant tuning voltage of the VCO, which is around 0.55 V. In order to keep the switches in the CPs in the saturation region, a maximum gate voltage of 1 V is needed to control them. A resistive potential divider can be employed to generate the required 1 V voltage for the switches. However, such a resistive divider needs to be small for fast switching and thus will consume a lot of DC power. A simple switched-capacitor driving stage, as shown in Fig. 7.7, is proposed to operate the switches in saturation. When the input is high, all the NMOS switches are turned on and discharge all the capacitors to zero. When the input is low, the PMOS switch is turned on and charges up the two capacitors in series. The ratio between the two capacitors is chosen to be 2 : 1, and as such, when 1.5 V is applied across the two capacitors, the output becomes 1 V. This capacitive divider can quickly switch to the required voltage from the reset state without consuming any DC power.

7.4.4. *Phase-frequency detector*

A traditional phase-frequency detector (PFD), as shown in Fig. 7.8, implemented by two D-type flip-flops and a NOR gate is used with some modification (Yoshizawa, Taniguchi and Nakashi, 1998). A slow NOR gate provides delay in the feedback

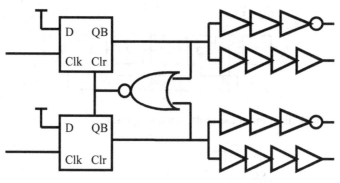

Fig. 7.8 Schematic of PFD

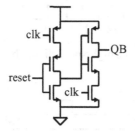

Fig. 7.9 Schematic of simplified TSPL D-type flip-flop

path to prevent a dead zone problem. A true single-phase logic (TSPL) D-type flip-flop is employed to simplify the design. Since the input of the flip-flop is always connected to high, the first stage of the TSPL flip-flop is omitted and simplified as Fig. 7.9. A chain of inverters is used to generate a pair of matched differential signals to drive the differential inputs of the current-steering charge pumps.

7.4.5. Prescaler

A prescalar with a division rate of 32, 32.5, . . . , 39.5 is used for the fractional-N synthesizer. The input frequency is around 900 MHz and the output is always 25.6 MHz when the loop is locked. As shown in Fig. 7.10, the prescalar is a cascade of a high-speed divide-by-2, 2.5, 3, 2.5 multi-modulus divider, two divide-by-2, three dual-modulus dividers and two divide-by-2 dividers. Based on the combinations of the internal status of the prescalar, the dividers are controlled to provide the overall division ratio from 32 to 39.5 with a step size of 0.5.

The high-speed multi-modulus divider is based on a phase-selection design (Craninckx and Steyaert, 1998; and Perrott, Tewksbury and Sodini, 1997). According to the timing diagram in Fig. 7.11, if an output of 90° phase lag is selected

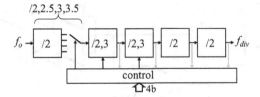

Fig. 7.10 Block diagram of the multi-modulus prescalar

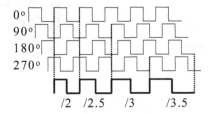

Fig. 7.11 Timing diagram of the high-speed divide-by-2, 2.5, 3, 3.5 multi-modulus divider

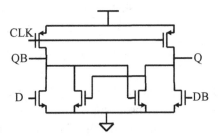

Fig. 7.12 Schematic diagram of the high speed D-latch

in the next cycle, the division ratio is increased by 0.5. As such, by selecting an output with the phase lag of 0°, 90°, 180°, or 270° in the next cycle, corresponding division ratios of 2, 2.5, 3, or 3.5, can be achieved.

The high-speed divide-by-2 divider, which is AC coupled from the output of the VCO, is a loop of two D-latches clocked by a pair of complementary signals. The high-speed D-latches, as shown in Fig. 7.12, have a full swing output and can provide a robust performance. Except for the clock inputs, all the transistors in the latch are NMOS without any stacking to increase the speed of operation.

7.4.6. *sigma–delta modulator*

A 10-bit digital third-order sigma–delta modulator is used to generate the outputs with the average value between 0 and 1 with a minimum resolution of $1/2^{10}$, and third-order high-pass quantization noise. The output of the sigma–delta modulator will control the multi-modulus prescalar. In this case, the multi-modulus prescalar

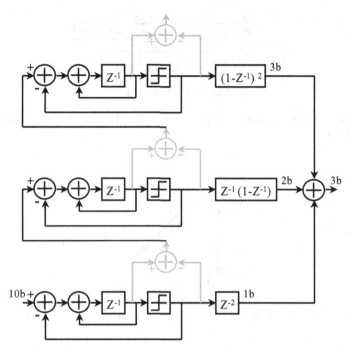

Fig. 7.13 System diagram of MASH-3 sigma–delta modulator

with a step size of 0.5 can have an average division ratio from 32 to 39.5 with a minimal step size of $0.5 \times 1/2^{10}$. For a 25.6 MHz reference frequency, the output frequency of the synthesizer can have a minimum frequency resolution of 25.6 MHz $\times$ $1/2^{11} = 12.5$ kHz.

In this design, a third-order cascade-type (MASH-3) modulator is used. It is a cascade of three first-order modulators with the quantization error of each stage being the input of the next stage, as shown in Fig. 7.13. The outputs of the three stages are passed through some filters and added together. The quantization noise at the outputs of the first two stages is effectively eliminated. The final combined output consists of a delayed version of the input and the third-order high-pass quantization noise. The schematic of the MASH-3 modulator is shown in Fig. 7.14. Three 10-bit digital accumulators are connected in series to realize the three first-order sigma–delta modulators and their carry-out outputs are passed to a quantization-noise-cancellation circuit for filtering and summing. The output of the MASH-3 modulator is a 3-bit output that represents the desired input fractional number with third-order high-pass quantization noise.

Since the sigma–delta modulator is used in a fractional-N frequency synthesizer, we usually want to have a constant DC input to provide a fixed division ratio between the reference frequency and the output frequency. However, the sigma–delta modulator will suffer from pattern noise if the input is silent or periodic. One

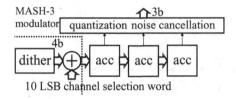

Fig. 7.14 Block diagram of the third-order digital sigma–delta modulator with a dither input

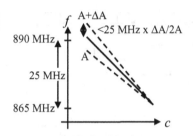

Fig. 7.15 Tuning curve by SCA with gain adjustment of finite resolution

effective way to solve this problem is to add some dither to the input to randomize it. Since the dither and the input signal appear together at the output, the dither should have the same noise shaping as the input signal, but a smaller amplitude in order not to affect the overall performance. As such, the dither has to be third-order high-pass as well. To generate the required dither, a pseudo-random generator realized with a 16-bit shift register is used, and the output is passed to a third-order digital high-pass filter. The 4-bit dither is then added to the lowest 4 LSBs of the sigma–delta modulator input. As verified by simulation and measurement, the dither effectively randomizes the input and removes the pattern noise.

7.4.7. Gain and offset adjustment for SCAs

The gain adjustment in SCAs is not perfect because of the finite resolution in digital multipliers. As shown in Fig. 7.15, the maximum frequency deviation due to the finite gain resolution is the product of the frequency range (25 MHz) and half of the minimum gain resolution.

By using a digital multiplier with a minimum resolution of $1/16$, the maximum frequency error due to the imperfect gain compensation is 781 kHz. To avoid a deviation of the value of the unit switchable capacitor in the array due to process variation, a 6-bit multiplier is used to provide the gain adjustment from $1/16$ to $3\frac{15}{16}$.

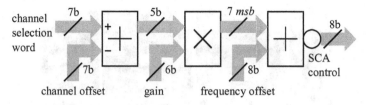

Fig. 7.16 Gain and offset adjustments for the SCA control

As the complexity of a digital multiplier is proportional to the product of the numbers of bits of the two inputs, an extra adder is included at the channel-selection word input, as shown in Fig. 7.16, to reduce its number of bits from 7 to 5. The first adder is used to convert the channel-selection word, which is also the input of the prescalar and the sigma–delta modulator, to the information of the channel number only (from channel 0 to channel 124). Since the SCA has a minimum resolution of about four channels, only groups of four channels (0–3, 4–7, . . .) will go into the gain adjustment. After the gain adjustment, an 8-bit frequency offset is added to adjust the center frequency within the 300 MHz tuning range. Finally, inverters are needed to invert all the signals to the SCA because the frequency is reduced as the capacitors are turned on.

7.5. Layout consideration

7.5.1. *Switchable-capacitor array*

The SCA includes 63 unit switchable capacitors and a half-size switchable capacitor. Each switchable capacitor includes a linear capacitor and a donut NMOS transistor as a switch to minimize the drain area and capacitance and thus to maximize the tuning range. The drain of the donut transistor is connected to the poly electrode of the linear capacitor. All the linear capacitors share a common n-well electrode in order to reduce the large redundancy and to improve the matching.

All the capacitors are laid out without sharp corners to minimize mismatches due to over-etching. The donut transistors and the linear capacitors have the same width of 5.4 μm. As a result, all the switchable-capacitors can be put side by side and form a compact layout as shown in Fig. 7.17. The length of each unit capacitor is minimized (~2.7 μm) to lessen the series resistance and capacitance thus maximizing the quality factor, Q. To reduce the mismatch between the unit switchable capacitors, they are arranged as shown in Fig. 7.18. All unit capacitors corresponding to a single bit are evenly distributed in the whole array. Moreover, all the bits, except the lowest two bits, are laid out to have a common centroid.

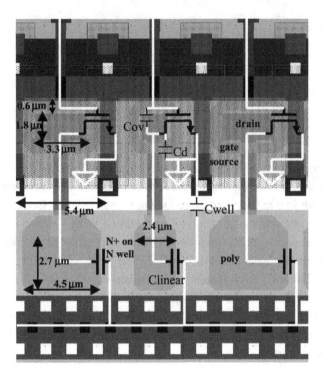

Fig. 7.17 Layout of SCA

7.5.2. Varactor layout

The varactor employs a parasitic *pn*-junction diode between the p_+ and *n*-well, as shown in Fig. 7.19. The two varactors used in the VCO are constructed using 18 and 186 *pn*-junction diodes in parallel, respectively. By minimizing the size of the unit diode and the distance between the p_+ electrode and n_+ ohmic contact, the series resistance of the diode, which depends on the length of the high-resistive *n*-well, is minimized and, as a consequence, the quality factor is maximized. For our case, a quality factor of around 30 is obtained for the varactors. For the same reason, octagonal unit diodes are drawn to keep a smaller distance (1.2 μm) between the p_+ and n_+ even at the corners.

7.5.3. Inductor layout

The inductors used are double-layer (Metal 3 and Metal 2) circular spiral inductors. Compared with a square spiral inductor with the same inductance, a circular spiral inductor has less series resistance and parasitic capacitance by a factor of $\pi/4$. A smaller serial resistance or a larger quality factor provides better phase-noise performance, and a smaller parasitic capacitance results in a larger available tuning

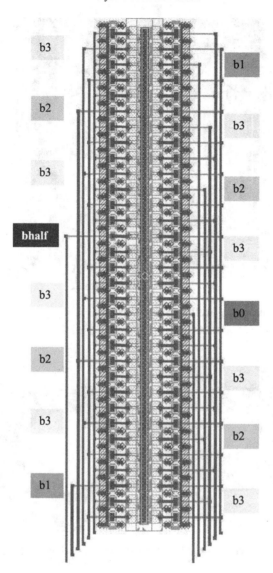

Fig. 7.18 Layout of half of the SCA

range for the VCO. Our previous measurements of testing structures show that single-layer and double-layer spiral inductors have a similar quality factor and parasitic capacitance. However, the double-layer spiral inductor occupies only about one quarter of the chip area.

The diameter of the spiral inductor is 280 μm with a hollow hole of 75 μm diameter. The spacing between spirals is minimal, 1.8 μm. Since the sheet resistances of the two metal layers are different (0.05 Ω/μm for Metal 3 and 0.07 Ω/μm for

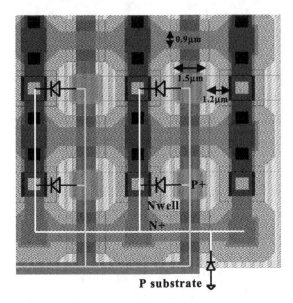

Fig. 7.19 Layout of p_+ n-well varactor

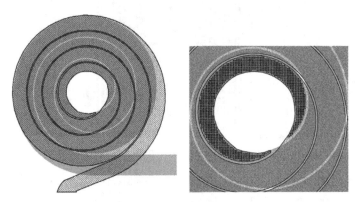

Fig. 7.20 Close-up view of the interconnection of two layers of the inductor

Metal 2), the widths and numbers of turns of the spirals of the two layers are designed to be different so that the same resistance per unit length is obtained, as shown in Fig. 7.20. As such, the total series resistance is not dominated by the spiral of either layer. The widths of the spirals in the Metal 2 and Metal 3 layers are 36 μm and 26 μm, respectively, while the numbers of turns are 2.5 and 3.5, respectively. The same resistance per unit length can still be maintained even if the width of the inner turn of the two spirals reduces. Such a special connection can maximize the size of the hollow hole to improve the quality factor and reduce the parasitic capacitance.

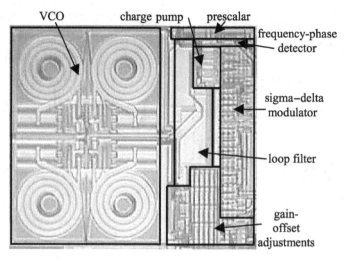

VCO   charge pump   prescalar

frequency-phase
detector

sigma–delta
modulator

loop filter

gain-
offset
adjustments

Fig. 7.21 Chip photo

The spiral inductors are designed and simulated using the ASITIC program. The inductance is 6.6 nH and the parallel parasitic capacitance is 0.6 pF. Owing to the large loss from the series resistance of the metals and the conductive substrate, the quality factor of the inductors is only around 2–2.5. It dominates the over-all quality factor of the LC tanks and limits the phase noise performance of the oscillator.

7.6. Experimental results

The prototype is fabricated by a HP 0.5 μm CMOS process with linear-capacitor and silicide-blocked options. The die photo is shown in Fig. 7.21, and the core area is $1.1 \times 0.9 \, \text{mm}^2$. The supply voltage is 1.5 V and power consumption is 30 mW. The output frequency range is measured from 857.6 MHz to 922.8 MHz with a minimal resolution of 25 kHz.

The SCA can provide monotonic frequency tuning from 760 MHz to 980 MHz with a gain of 1 MHz per step at 900 MHz as shown in Fig. 7.22. The differential non-linearity (DNL) of the SCA is much less than 0.5 LSB. As such, it is possible to use an SCA with a larger number of bits to provide a finer frequency step and a higher resolution. The only limitations are the minimal feature sizes of the capacitors and the switches, and the additional parasitic capacitance.

The measured gains of the VCO are smaller than expected at only 7 MHz/V and 0.7 MHz/V. The measured tuning voltage of the varactors as a function of the channel number is shown in Fig. 7.23. The maximum variation of the tuning voltage is 0.24 V. It is due, (a) to the finite frequency resolution, 1 MHz, provided by the SCA, (b) to the non-linearity of the $f = 1/\sqrt{LC}$ relation, 300 kHz, (c) to the

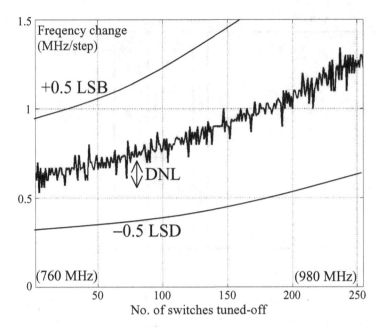

Fig. 7.22 Frequency change plotted against number of switches tuned-off

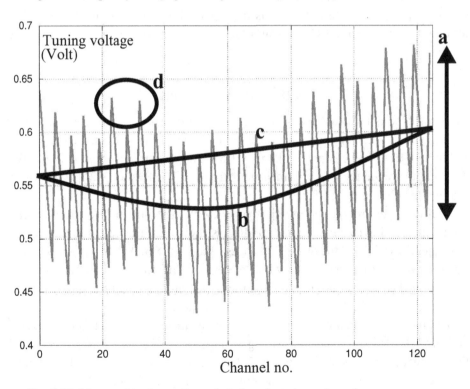

Fig. 7.23 Measured tuning voltage plotted against channel number

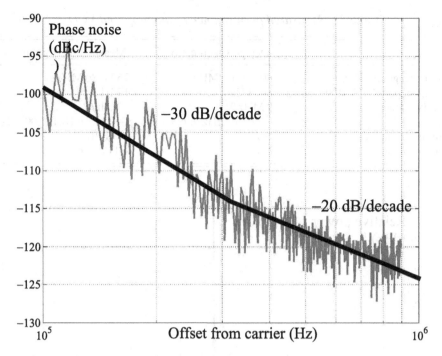

Fig. 7.24 Measured output phase noise of the synthesizer

250 kHz error caused by the imperfect gain adjustment provided by the digital multiplier with a finite number of bits, and d) to the differential non-linearity of the SCA.

As shown in Fig. 7.24, the phase-noise measurements of the synthesizer at offsets of 400 kHz and 600 kHz are −116 dBc/Hz and −118 dBc/Hz, respectively. Since the frequencies are out of the loop bandwidth, they are mainly determined by the phase noise of the free-running VCO. However, the phase noise of the VCO is limited by the poor quality of the spiral inductors. Within a 200 kHz offset, the phase noise with the slope −30 dBc/decade is dominated by the flicker noise of the transistors in the VCO.

Despite the high tolerance of substrate noise, the reference frequency of 25.6 MHz can still couple to the VCO, and spurs of −67 dBc at 25.6 MHz off-set are observed as shown in Fig. 7.25. The coupling is due to the short separation (∼200 μm) in the dense layout between the circuits that are clocked by the reference frequency and the VCO.

The measured loop bandwidth is around 80 kHz, estimated from the step response. In the case of the maximal change in the tuning voltage of the varactors, the settling time, including the switching of the SCA, is measured to be less than 100 μs for the output frequency to be within 20 kHz of the final value, as shown in Fig. 7.26.

Table 7.1 *Measured performance of the proposed VCO*

	Design specifications	Measured performances
Tuning range by SCA	800–1100 MHz	760–980 MHz
Tuning step by SCA at 890 MHz	800 kHz	1 MHz
Gain by varactor	9 MHz/V	7 MHz/V
Phase noise	−119 dBc/Hz at 600 kHz	−118 dBc/Hz at 600 kHz
Amplitude matching	N.A.	<0.1 dB
Phase matching	N.A.	<1 degree

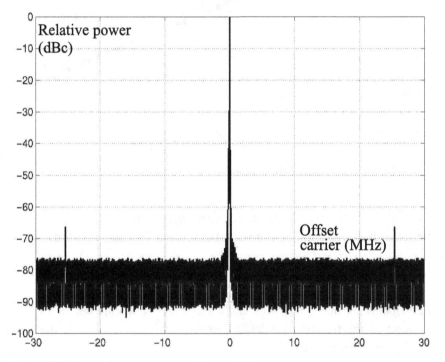

Fig. 7.25 Measured output spurs of the synthesizer

7.7. Performance summary and evaluation

Summaries of the performance of the VCO and the frequency synthesizer are shown in Table 7.1 and Table 7.2, respectively. Owing to the large capacitance in the SCA, the measured center frequency, tuning range and the gain of the VCO are smaller than designed for, while the frequency-tuning step is larger. The measured phase-noise performance is 1 dB lower than the design value because of the unexpected poor quality factor of the inductors. Most of the measured performances of the

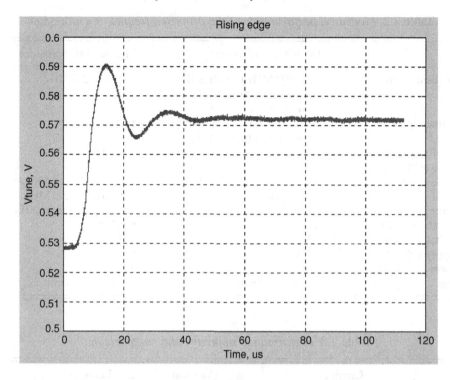

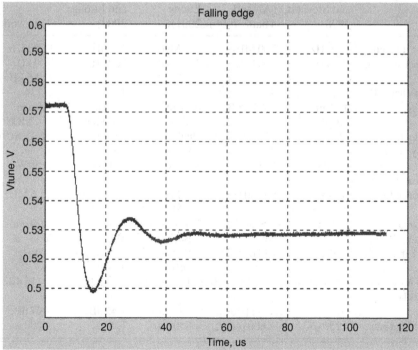

Fig. 7.26 Measured tuning voltage (for rising and falling step input)

Table 7.2 *Measured performance of the proposed frequency synthesizer*

	Design specifications	Measured performance
Frequency range	865–890 MHz (70 MHz IF receiver) 890–915 MHz (transmitter)	844.8–972.8 MHz
Frequency resolution	200 kHz (or finer for digital modulated transmitter)	12.5 kHz
Amplitude matching	N.A.	<0.1 dB
Phase matching	N.A.	<1°
Phase noise	<−119 dBc/Hz at 600 kHz	−118 dBc/Hz at 600 kHz
Spurs	<−88 dBc	−67 dBc at 25.6 MHz
Settling time	<865 μs	<200 μs to within 20 kHz
Loop bandwidth	80 kHz	80 kHz
Digital modulation	GMSK	FSK
Supply voltage	<2 V	1.5 V
Power consumption	<50 mW	30 mW
Area	<2 mm × 2 mm	1.1 mm × 0.9 mm

Table 7.3 *Performance summary and comparison*

	(Craninckx and Steyaert, 1998)	(Ali and Tham, 1996)	(Yan and Luong, 2001)	(Kan, Leung and Luong, 2002)	This work
Center frequency	1.8 GHz	900 MHz	900 MHz	1.8 GHz	900 MHz
Channel spacing	200 kHz	600 kHz	200 kHz	200 kHz	200 kHz (12.5 kHz)
No. of channels	124	41	124	103	>124
Process	0.4 μm CMOS	25 GHz BJT	0.5 μm CMOS	0.5 μm CMOS	0.5 μm CMOS
Architecture	FN	FN	Dual loop	Dual loop	FN & SCA
Supply voltage	3 V	2.7–5 V	2 V	2 V	1.5 V
Power consumption	51 mW	50 mW	34 mW	95 mW	30 mW
Reference freq.	26.6 MHz	9.6 MHz	205 MHz 1.6 MHz	100 MHz 800 kHz	25.6 MHz
Chip area	3.23 mm^2	5.5 mm^2	2.64 mm^2	2 mm^2	0.99 mm^2
On-chip filter	Yes	No	Yes	Yes	Yes
Loop bandwidth	45 kHz	4 kHz	40 kHz 27 kHz	120 kHz 42 kHz	80 kHz
Phase noise at 600 kHz	−121 dBc/Hz	−117 dBc/Hz	−121 dBc/Hz	−111 dBc/Hz	−118 dBc/Hz
Spurs	−75 dBc	<−110 dBc	−79.5 dBc	−45 dBc	−67 dBc
Switching time	<250 μs	<600 μs	<830 μs	128 μs	<200 μs

frequency synthesizer meet the design specifications, including the frequency range, frequency resolution, settling time, loop bandwidth, supply voltage, power consumption and chip area. The spurs, which cannot be predicted, are quite large because of the aggressive layout. However, due to the carefully chosen reference frequency of 25.6 MHz, the spurs will not pose a serious problem in the receiver system.

Table 7.3 summarizes the measured performance of the proposed synthesizer in comparison with some other designs reported recently. Other designs are based on fractional-N or dual-loop architecture, while the proposed design is based on fractional-N architecture and SCA tuning. Compared with other designs, the proposed synthesizer design achieves the lowest supply voltage, the lowest power consumption, the smallest chip area, and the highest loop bandwidth. The phase noise and switching time are similar to the others. Only the spurs are higher than some designs, due to the smaller chip area and denser layout.

8

A 1 V 5.2 GHz fully integrated CMOS synthesizer for WLAN IEEE 802.11a

A very low voltage synthesizer designed for WLAN is introduced in this chapter. Novel circuit designs discussed in Chapter 5 are demonstrated, such as for the VCO and programmable divider. The measured phase noise is $-136\,\mathrm{dBc/Hz}$ at an offset of 20 MHz from the carrier, while the spurious tone is better than $-80\,\mathrm{dBc}$ at the offset of the reference clock frequency. The synthesizer dissipates 27.5 mW from a single 1 V supply. The total core area occupies $1.03\,\mathrm{mm}^2$.

8.1. WLAN overview

IEEE 802.11a is a new standard aimed at high-speed wireless LAN communication systems. A high data rate of up to 54 Mbps can be offered between portable devices. It is designed for the 5 GHz frequency spectrum employing orthogonal frequency division multiplexing (OFDM) as its modulation scheme.

IEEE 802.11a is allocated a bandwidth of 300 MHz, which is separated into three bands with 100 MHz for each, according to different transmitting powers (Fig. 8.1). The three bands are the lower band, the middle band and the upper band. The lower band and middle band are specified for indoor communication, and the maximum transmit powers are 40 mW and 200 mW, respectively. The upper band is suited for outdoor use with a larger transmitting power of up to 800 mW. Each band contains four channels and each channel occupies 20 MHz bandwidth with 48 data sub-carriers, four pilot sub-carriers, and one nulled sub-carrier. Another 11 sub-carriers are discarded during signal processing (Agilent, 2003).

In an indoor environment, transmit signals are easily reflected from different objects and, therefore, the same transmitted signals can arrive at the receiver at different times. OFDM is a new modulation scheme which is able to minimize this multipath effect. It offers the capability of transmitting a numbers of carriers at the same time instead of transmitting one carrier every time.

Table 8.1 *Frequency allocation of IEEE 802.11a*

Frequency band	Center frequency (/GHz)
UN-II* lower band (5.15–5.25 GHz)	5.18, 5.20, 5.22, 2.24
UN-II middle band (5.25–5.35 GHz)	5.26, 5.28, 5.30, 5.32
UN-II upper band (5.725–5.825 GHz)	5.745, 5.765, 5.785, 5.805

* Unlicensed National Information Infrastructure

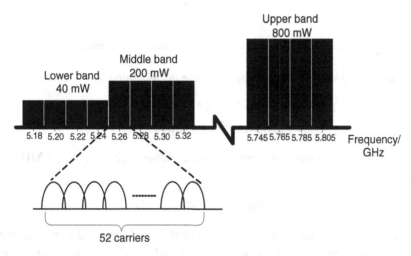

Fig. 8.1 Frequency allocation of IEEE 802.11a

The frequency spectrum is illustrated graphically in Fig. 8.1. From the first channel of the lower band to the last channel of the middle band, the center frequencies are from 5.18 GHz to 5.32 GHz. For the upper band, the center frequency ranges from 5.745 GHz to 5.805 GHz. Table 8.1 summarize the frequency allocation.

8.2. Design specification

The phase noise performance of a synthesizer is significant to the whole system adopting OFDM as modulation. The specification can be determined by the unwanted down-conversion of the adjacent channel interferences. Considering that the receiver achieves the highest data rate, which is 54 Mbps with a minimum sensitivity of −65 dBm, as shown in Table 8.2, and an adjacent interferer is 40 dB stronger than the desired channel, while the SNR is assumed to be 19 dB for a BER

Table 8.2 *IEEE 802.11a receiver performance*
requirements

Data rate (Mbps)	Minimum sensitivity (dBm)
6	−82
9	−81
12	−79
18	−77
24	−74
36	−70
48	−66
54	−65

of 10^{-6} in a 64-QAM system, the phase noise is calculated as

$$L(20\,\text{MHz}) \leq -40 - 19 - 10\,\log(20 \times 10^6)$$
$$L(20\,\text{MHz}) \leq -132\,\text{dBc/Hz}. \tag{8.1}$$

By assuming a $1/f_2$ noise spectrum, the phase noise requirement at 1 MHz offset from the carrier is

$$L(1\,\text{MHz}) \leq -107\,\text{dBc/Hz}. \tag{8.2}$$

Similarly, the spurious tone requirement can be derived based on the specification of the desired signal power, blocking signal power and signal-to-noise requirement for the receiver. The blocking signal is as large as $-30\,\text{dBm}$. Thus, it can be calculated as

$$S_{spur} \leq S_{desired} - S_{block} - \text{SNR} \tag{8.3}$$
$$\leq -65 - (-30) - 19$$
$$\leq -54\,\text{dB}.$$

The frequency synthesizer is aimed at indoor communication with the frequency range designed from 5.15 GHz to 5.35 GHz, including both the lower and middle bands.

8.3. Synthesizer architecture

As discussed in Section 2.5, there are different synthesizer architectures such as integer-N, fractional-N or dual-loop. Based on the fact that the channel spacing of IEEE 802.11a is relatively wide compared with other standards like GSM, the division ratio can be small with a high reference frequency. In addition, integer-N is relatively simple compared with the other two topologies, and thus it is adopted for

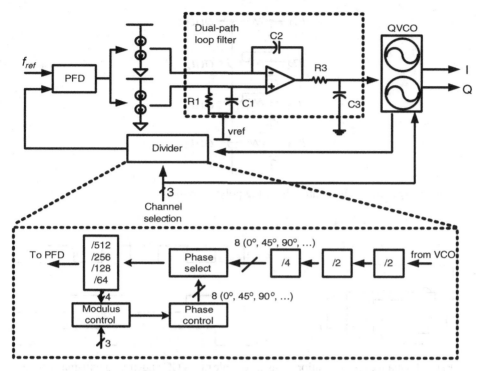

Fig. 8.2 Block diagram of the proposed prescaler and synthesizer

this design. Nevertheless, this synthesizer is specified to operate with a 1 V supply, so some high frequency building blocks, such as the VCO and the dividers should be designed carefully to prevent failure when the voltage supply varies.

Figure 8.2 shows the system architecture of a synthesizer designed for WLAN 802.11a transceiver systems with a data rate as high as 54 Mbps (Leung and Luong, 2000b). The system comprises a quadrature voltage controlled oscillator (QVCO) for generating in-phase and quadrature-phase, a programmable divider, a PFD, two CPs, and a third-order dual-path loop filter. The division ratio is 498 : 512 in steps of two with a reference frequency of 10 MHz and a frequency step of 20 MHz. A novel prescaler design is proposed based on the phase-switching technique which relaxes the speed requirement under a low-voltage supply (Craninckx, 1996). Three digital bits are implemented to simultaneously select the division ratio in the prescaler and provide coarse frequency tuning to the QVCO to select channels in both the lower band (5.15–5.25 GHz) and the middle band (5.25–5.35 GHz). As a consequence, the QVCO can be designed with a small VCO gain (K_{vco}) to ensure good performance with a low supply voltage, not only in terms of phase noise and spur performance, but also in terms of on-chip capacitors and chip area for the loop filter.

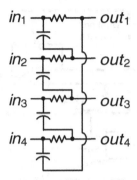

Fig. 8.3 Quadrature generation by using polyphase filter

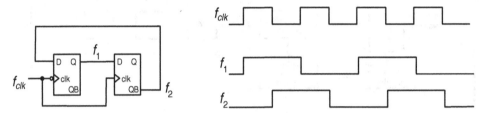

Fig. 8.4 Quadrature generation by using double rate VCO and divider with master–slave configuration

8.4. Quadrature phase generation

Quadrature phases in which two signals are 90° out of phase are often used for an image-rejected receiver. A polyphase filter, or a RC-CR phase shifter, as shown in Fig. 8.3, is widely used to generate quadrature signals from a single-phase VCO (Pache *et al.*, 1995; Behbahani *et al.*, 2000). Nevertheless, the passive network, which comprised resistors and capacitors to generate poles and zeros, is relatively dependent on the operating frequency. A higher-order approach of cascading several stages of polyphase filter has been demonstrated by Behbahani *et al.* (2001) with the intention of extending the image-rejection bandwidth. However, the oscillating amplitude and noise suffer unavoidably.

Another approach to generating quadrature clock phases is to use a divide-by-2 circuit. As depicted in Fig. 8.4, a single clock phase VCO signal drives the SCL divider with a master–slave configuration. The VCO frequency is designed to be double the required frequency in order to acquire a divider output of the desired frequency with quadrature phases. Obviously, this method needs to dissipate more power because of the double rate implementation of both the VCO and the high-speed dividers. This is especially difficult under low-voltage operation.

Table 8.3 *Component parameters used for the synthesizer*

Component parameter	Value
Division ratio N	512
Loop bandwidth ω_c	$2\pi \times 50\,\text{kHz}$
I_{CP}	$1\,\mu\text{A}$
Charge pump current ratio B	30
R_p	$23\,\text{k}\Omega$
C_p	$25.6\,\text{pF}$
C_z	$20.5\,\text{pF}$
R_4	$23\,\text{k}\Omega$
C_4	$25.6\,\text{pF}$
K_v	$100\,\text{MHz/V}$

To facilitate the implementation of low power and the frequency independence of quadrature phase generation, two identical cross-coupled VCOs can be utilized as mentioned in Chapter 3, but with the trade-off of increasing chip area.

8.5. Behavioral simulation

The parameters used for the synthesizer are shown in Table 8.3. By using these parameters for stability analysis, the phase margin of the synthesizer system is 60° with the cross-over frequency at 50 kHz as illustrated in Fig. 8.5. By using SpectreRF, the transient response is depicted in Fig. 8.6. With a division ratio of 512 and the reference clock operating at 10 MHz, the output frequency becomes 5.12 GHz as shown in Fig. 8.7.

Figure 8.8 shows the phase noise performance of the synthesizer. At 1 MHz offset from the carrier frequency, the phase noise at the output of the synthesizer depends on the noise of the VCO, which is about −113 dBc/Hz.

8.6. Circuit implementation

8.6.1. Programmable frequency divider

As shown in Fig. 8.2, the proposed programmable frequency divider is implemented based on the phase-switching approach. This lessens the need for low-voltage digital circuits and minimizes the power consumption because only one divide-by-2 circuit needs to operate at the highest frequency. Successive asynchronous dividers work at lower frequencies and can be designed with smaller device sizes and lower power consumption.

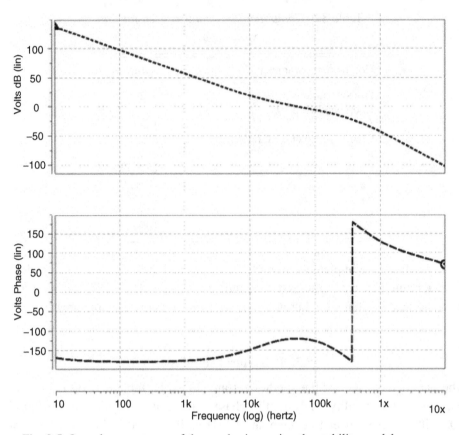

Fig. 8.5 Open-loop response of the synthesizer using the stability model

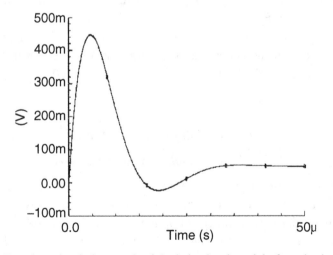

Fig. 8.6 Transient simulation result of the behavioral model of synthesizer

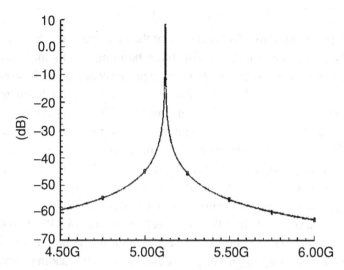

Fig. 8.7 Output spectrum of the synthesizer using the behavioral model

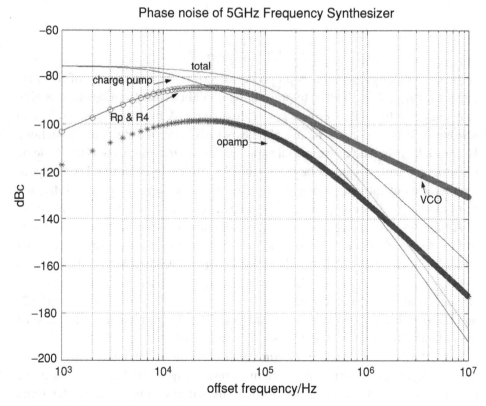

Fig. 8.8 Phase noise plot of the synthesizer

8.6.1.1. Divide-by-2

There are two main reasons why the design of the first frequency divider is the most critical and challenging compared with other building blocks in the synthesizer. Firstly, the divider needs to be able to operate properly at the synthesizer's highest frequency (5.2 GHz). Secondly, the divider needs to have an input frequency range at least as wide as the frequency tuning range of the QVCO to cover all the channels in the presence of process and temperature variation (>500 MHz). To realize a frequency divider with such stringent performance under low supply voltages and low power consumption, a divider architecture using D-latches in a master–slave configuration is used (Fig. 4.2). The device sizes are scaled for operating in the 6 GHz range. Such a design margin guarantees that the divider is able to work under process variation of the QVCO as well as the divider itself. As the clock signals are from the QVCO output, whose DC point is around the supply voltage, level shifters are required to properly bias the gates of the loading transistors and of the current sources. The divider can overcome the speed limitation problem and achieve high performance with a 1 V supply voltage without using a voltage doubler.

8.6.1.2. Divide-by-4

A phase-switching operation is proposed to perform at a frequency of $f_{VCO}/16$, rather than $f_{VCO}/8$, or higher, to prevent the digital circuits from working at high frequency and to minimize their power dissipation. A backward phase selection scheme is also used for avoiding glitches which occur during phase switching. However, by using the conventional approach, the division step that is available by the programmable divider is larger (Craninckx, 1996). In order to relax the speed requirements of the phase-selection circuits and to achieve the required division resolution, the divider outputs need to be designed to provide finer clock phases.

   An approach to using two divide-by-2 circuits working in parallel is shown in Fig. 8.9(a) (Shu *et al.*, 2003). The total number of output patterns of Divider 2a and Divider 2b is eight but there are two possible patterns as illustrated in Fig. 8.9(b). The reason why there are two patterns is that the initial outputs of Divider 2a and Divider 2b can be either one or zero (Fig. 8.10). Because Divider 2a and Divider 2b synchronize the input of Divider 1 only, the phase relation between the *I*-channel of Divider 2a and Divider 2b can be either 45° or −135°.

   Since the direction of phase selection determines the output phase from P7 to P0 and then repeats, the two possible patterns shown in Fig. 8.9(b) lead to wrong phase switching and result in a wrong division ratio from the programmable divider. Extra circuits are therefore needed to detect such random behavior.

   The configuration of the high-frequency dividers, together with the phase switching circuits, in the proposed programmable divider is shown in Fig. 8.2. As the phase

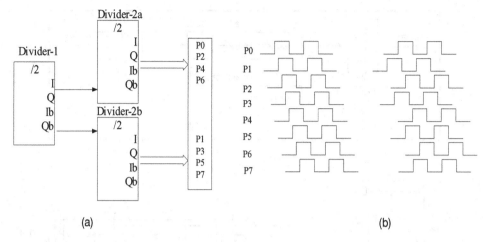

Fig. 8.9 (a) Dividers in parallel configuration with eight output clock phases (b) two possible output patterns of the eight clock phases

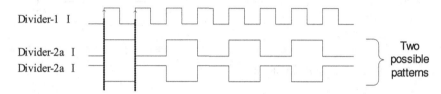

Fig. 8.10 Two possible output patterns of dividers in parallel configuration due to different initial status

switching circuit is designed to operate at $f_{VCO}/16$, a divide-by-4 circuit is proposed, which also operates at a frequency of $f_{VCO}/16$. The divide-by-4 used in the synthesizer has advantages over the conventional approach of a cascading divide-by-2. Firstly, if a parallel configuration is employed, one extra divider operating at $f_{VCO}/8$ is required. The proposed approach can save the divider operating at $f_{VCO}/8$ and save area. Secondly, the divide-by-4 circuit has only one phase pattern, and operates as a ring oscillator while it is injection-locked by the input signal of the previous stage divider. As a result, this injection-locked oscillator can oscillate at a frequency of $f_{VCO}/16$ and has outputs with only one phase pattern. In other words, there is no extra phase detection circuit required. The divide-by-4 circuit is realised by cascading four D-latches, as shown in Fig. 8.11, to generate eight clock phases that are 45° out of phase with each other. The total phase shift is 360° for oscillation. The current source M1 converts the input voltage of CLK to current, and passes through the input devices M3 and M4. The negative-g_m cell formed by M5 and M6 is to keep the value of output once CLKB is high. The current is finally converted to an output voltage by the loads M7 and M8. Because the current is synchronized with the input signal CLK, the output value toggles based on the input values D and

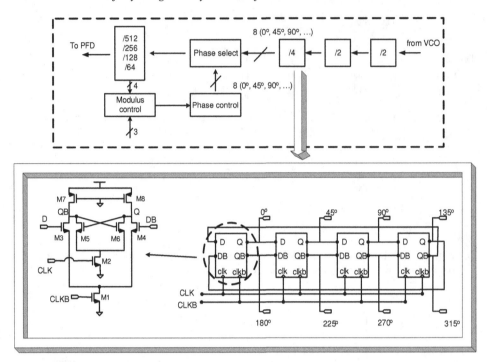

Fig. 8.11 Schematic of divide-by-4

DB when the CLK signal is active, as shown in Fig. 8.12. As such, a divide-by-4 function can be implemented.

8.6.1.3. *Phase switching circuits*

Owing to the delay uncertainty of the flip-flops, glitches can happen during the phase selection, which could result in a wrong division ratio (Craninckx, 1996). A backward phase-selection scheme can solve the problem of glitches during the phase selection, but it requires the circuits to work at higher frequencies during the phase transitions in addition to careful design of the switching control (Shu *et al.*, 2003). The proposed phase-selection circuitry, as illustrated in Fig. 8.13, uses an 8-bit shift register to control the phase-selection circuit with only one of the input phases from the divide-by-4 selected. No glitches occur during transition. A single-stage 8-to-1 multiplexer is used to shorten the delay from the input to the output.

Even if the phase selection suffers from a finite rising time due to low-voltage operation, the phase can be switched successfully and correctly as long as the current stage of the phase-selection circuit does not turn off before the next stage is completely on. In the proposed design, shown in Fig. 8.13, the sensing PMOS transistor is included to ensure that this is the case. Simulation results verify that, without the sensing transistor, the output of the phase-selection circuitry suffers

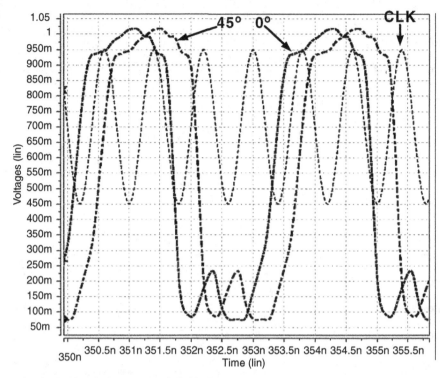

Fig. 8.12 Simulation result of divide-by-4

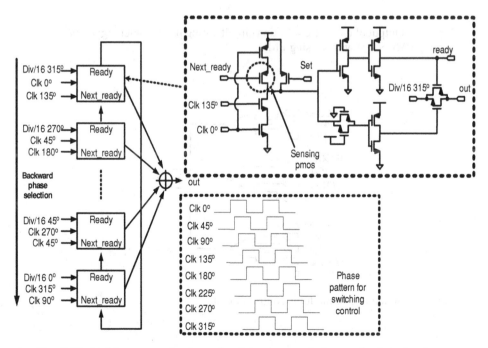

Fig. 8.13 Architecture and schematic of the proposed phase-selection circuitry

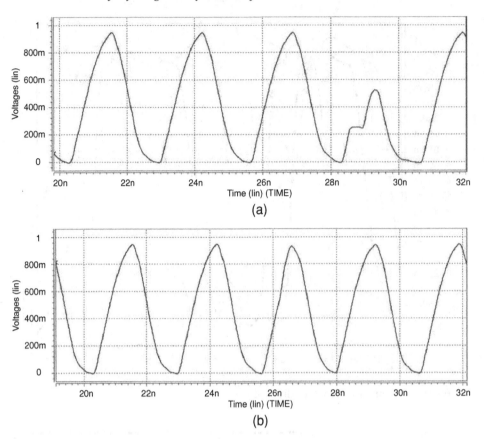

Fig. 8.14 Output of the phase selection circuit during phase switching: (a) without sensing PMOS, (b) with sensing PMOS

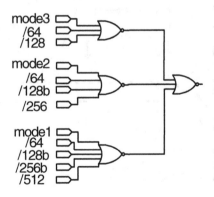

Fig. 8.15 Schematic of modulus control circuits

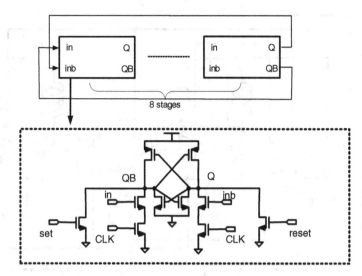

Fig. 8.16 Phase control schematic

from a serious problem of an unacceptably low swing during the phase switching, as shown in Fig. 8.14(a). The swing is limited because of the finite rising time and because of the intrinsic delay of the digital circuits under low-voltage supply. The problem is completely remedied, as illustrated in Fig. 8.14(b), by adding the sensing PMOS without increasing power consumption.

8.6.1.4. Modulus control circuits and phase control circuits

The modulus control circuits are constructed by using combination circuits of NOR gates. Figure 8.15 shows the schematic of the control circuits. While the phase control circuits comprise eight stages of D-latches, as shown in Fig. 8.16, and the last stage output is connected to the first stage input with cross-coupled configuration, the phase shift depends on the output of the modulus control circuit.

8.6.2. Quadrature LC oscillator

Figure 8.17 shows the schematic of the proposed quadrature LC oscillator (QVCO). It consists of two identical LC VCOs with negative-transconductance cells M_{1a-2a} and M_{1b-2b} to compensate for the losses in the LC tanks. The LC VCOs are directly coupled and cross-coupled by M_{3a-4a} and M_{3b-4b} to achieve four outputs that are $90°$ out of phase.

Owing to the low supply and control voltages, the pn-junction and MOS capacitance are not suitable for frequency tuning over the two bands. In the proposed design, the frequency tuning is achieved by both coarse tuning and fine tuning. As

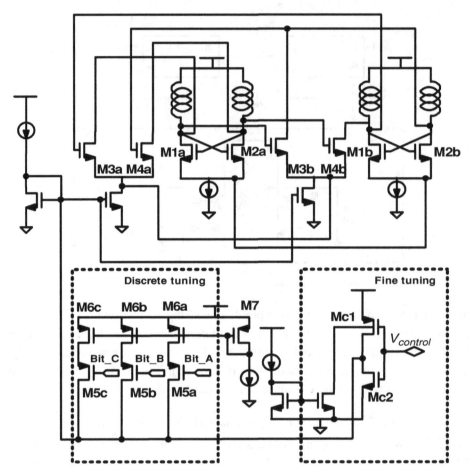

Fig. 8.17 Schematic of QVCO

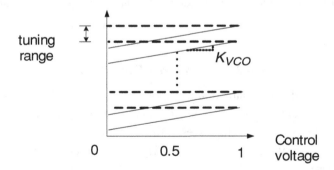

Fig. 8.18 QVCO tuning combining with fine and coarse tuning for low noise implementation

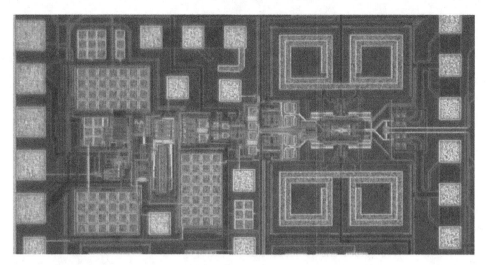

Fig. 8.19 Chip photo of the proposed 5.2 GHz Synthesizer for WLAN

mentioned earlier, the coarse frequency tuning is employed to minimize the VCO gain by using three binary-weighted channel-selection bits together with the channel-selection transistors M_{5a-c} and M_{6a-c}, as shown in Fig. 8.17. As a result, the K_{VCO} is small, while the total tuning range can still cover the required band, as shown in Fig. 8.18. This is beneficial to the noise and spur performance of the synthesizer. The offset adjustment for the discrete frequency step can be calibrated by varying the current source M7. The fine frequency tuning is obtained by varying the tranconductances of the coupling transistors M_{3a-4a} and M_{3b-4b}.

8.6.3. Loop filter, CPs and PFD

The loop filter is constructed based on a dual-path loop filter to reduce the on-chip capacitance to about 70 pF and thus minimize the chip area. The filter in the locked state keeps the outputs of the CPs at the same potential to minimize charge sharing and to reduce spurious tones at the output of the synthesizer. The complementary outputs of UP and UPB, DOWN and DOWNB generated by the PFD are designed to have some finite rise and fall time so that it can maintain a match of the sink and source current that flows into the loop filter.

8.7. Experimental results

8.7.1. Introduction

The proposed synthesizer has been fabricated in a 0.18 μm six-metal CMOS process. Figure 8.19 shows the microphotograph of the fabricated chip. The

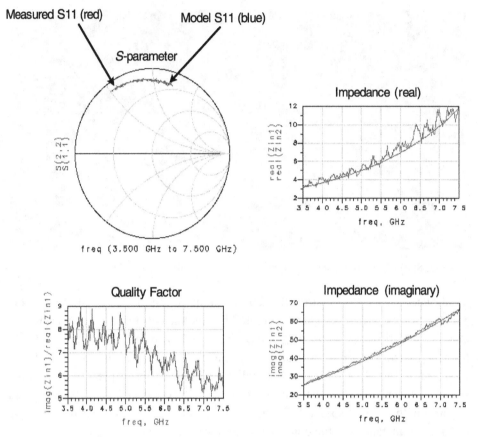

Fig. 8.20 Measured results of the on-chip inductance for the WLAN synthesizer

core chip area is $1.03\,\text{mm}^2$. The total power consumption, with a 1 V supply, is $27.5\,\text{mW}$.

8.7.2. Inductor measurement

Figure 8.20 shows the measured quality factor of the on-chip inductors. This is about 7 at $5.2\,\text{GHz}$, and the measured inductance is 1 nH.

8.7.3. Measurement of the QVCO

The measured tuning range of the QVCO is plotted in Fig. 8.21. Together with the coarse tuning by the three channel-selection bits, the QVCO achieves a total frequency tuning range of around $200\,\text{MHz}$ with the gain K_{VCO} being around $75\,\text{MHz/V}$. As expected, for each setting of the digital bits, the tuning range of

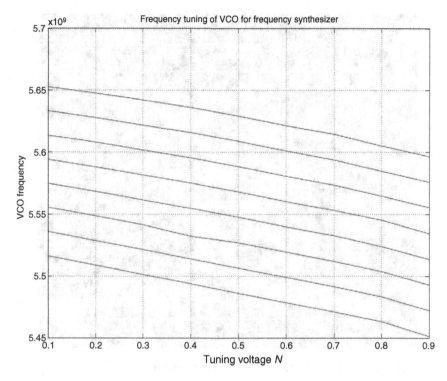

Fig. 8.21 Measured tuning range of the QVCO

the QVCO is quite linear over the whole operation region of the dual-path loop filter.

8.7.4. Measurement of the synthesizer

The center frequency of the QVCO is shifted up due to an inaccurate model and overestimation of the parasitic capacitance. As a result, the reference is set at 11 MHz for all the close-loop measurements. With the robustness of the proposed architecture for the programmable divider, the synthesizer works properly even with a supply voltage as low as 0.85 V. The output spectrum of the QVCO in the locked state is shown in Fig. 8.22. To verify the functionality of the programmable divider, the division ratio is set to be 498. The output frequency is measured to be 5.478 GHz with a reference clock of 11 MHz. The prescaler works properly without any glitch problems. From the output spectrum, the spur performance is better than −80 dBc at a 11 MHz offset from the carrier.

The phase-noise performance of the frequency synthesizer is shown in Fig. 8.23. The measured phase noise at a 20 MHz offset is −136 dBc/Hz. The in-band phase noise is around −65 dBc/Hz, which is mainly contributed by the phase noise of

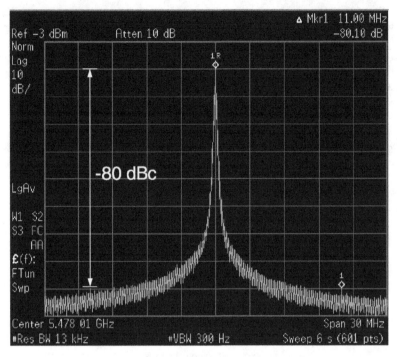

Fig. 8.22 Measured output spectrum of the synthesizer

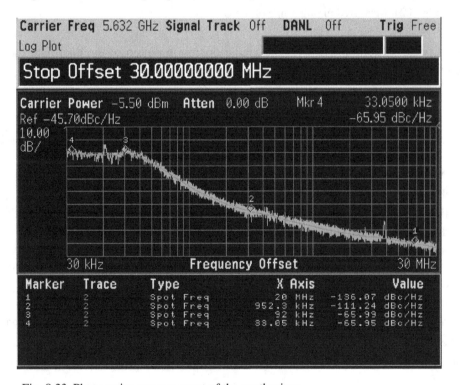

Fig. 8.23 Phase-noise measurement of the synthesizer

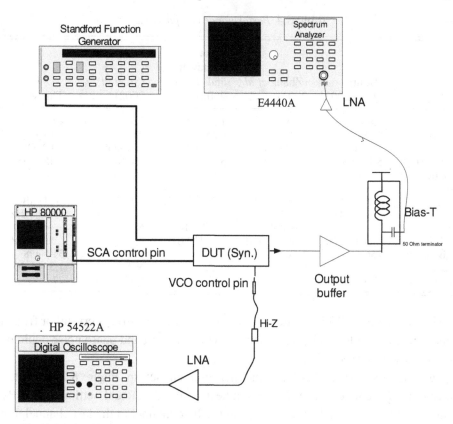

Fig. 8.24 Settling time measurement setup

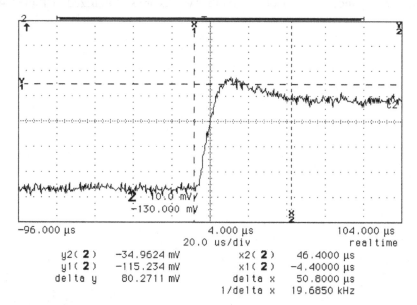

Fig. 8.25 Measured result of settling time of the synthesizer

Table 8.4 *Performance comparisons*

	(Bouras et al., 2003)	(Zhang et al., 2003)	(Lee, Samavati and Rategh, 2002)	(Su et al., 2002)	This work
Supply	1.8 V	1.8 V	1.8 V	2.5 V	1 V
Process	0.18 μm CMOS	0.18 μm CMOS	0.25 μm CMOS	0.25 μm CMOS	0.18 μm CMOS
Freq. (GHz)	5.15–5.825	5.15–5.35	5	4.128–4.272	5.45–5.65
Quadrature output	*RC–CR*	Yes	Yes	*RC–CR*	Yes
Phase Noise	−115 dBc/Hz at 1 MHz	−113 dBc/Hz at 1 MHz	−107 dBc/Hz at 1 MHz	−112 dBc/Hz at 1 MHz	−111 dBc/Hz at 1 MHz
Spurious tone	<−65 dBc at 10 MHz	<−66 dBc at 13.3 MHz	<−70 dBc at 11 MHz	N.A.	<−80 dBc at 11 MHz
Division ratio	256	N.A.	512	512	512–498
Size (mm^2)	N.A.	N.A.	N.A.	N.A.	1.3 × 0.76
Power (mW)	N.A.	56	25.3	180	27.5

the signal generator. With a better reference signal, either from a crystal or from a signal generator followed by a frequency divider, the in-band phase noise could be significantly improved.

Figure 8.24 shows the settling-time measurement set-up. Figure 8.25 plots the measured transient response of the synthesizer for the actual settling time. By switching the channel selection pins from the first channel to the last channel, the settling time is about 51 μs.

The performance comparisons with other works are summarized in Table 8.4.

References

Adler, R. (1946). A study of locking phenomena in oscillators. *Proceedings of the IRE and Waves and Electrons*, **34**, 351–357, June.

Agilent (2003). *Application Note of Marketing 802.11G Transmitter Measurements*, Agilent Technologies, Feb.

Akazawa, Y., Kikuchi, H., Iwata, A., Matsuura, T. and Takahashi, T. (1983). Low power 1 GHz frequency synthesizer LSIs. *IEEE Journal of Solid-state Circuits*, **18**, 115–121, Feb.

Ali, A. and Tham, J. L. (1996). A 900 MHz frequency synthesizer with integrated LC voltage-controlled oscillator. *Proceedings of IEEE International Solid-state Circuits Conference 1996*, 390–1.

Andreani, P. and Mattisson, S. (2000). On the use of MOS varactors in RF VCOs. *IEEE Journal of Solid-state Circuits*, **35**, 905–910, June.

Andreani, P. (2001). A 1.8 GHz monolithic CMOS VCO tuned by an inductive varactor. *IEEE International Symposium on Circuits and Systems*, 714–717, May.

Aytur, T. and Khoury, J. (1997). Advantages of dual-loop frequency synthesizers for GSM applications. *IEEE International Symposium on Circuits and Systems*, June.

Behbahani, K. F. and Abidi, A. (1998). RF CMOS oscillators with switched tuning. *Proceedings of the IEEE 1998 CICC*, pp. 555–8.

Behbahani, F., Leete, J., Kishigami, Y. *et al.* (2000). A 2.4 GHz low-IF receiver for wideband WLAN in 6 μm CMOS-architecture and front-end. *IEEE Journal of Solid-state Circuits*, **35**, 1908–1916, Dec.

Behbahani, F., Kishigami, Y., Leete, J. and Abidi, A. (2001). CMOS mixers and polyphase filters for large image rejection. *IEEE Journal of Solid-state Circuits*, **36**, 873–887, June.

Best, R. (1999). *Phase-locked Loops: Design, Simulation, and Applications*, McGraw-Hill.

Bouras, I. *et al.* (2003). A digitally calibrated 5.15–5.825 GHz transceiver for 802.11a wireless LANs in 0.18 mm CMOS. *IEEE International Solid-state Circuits Conference*, pp. 352–353, Feb.

Buchwald, A. W. and Martin, K. W. (1991). High-speed voltage-controlled oscillator with quadrature outputs. *Electronics Letters*, **27**, 309–310, Feb.

Cadence. (1988). *Oscillator Noise Analysis in SpectreRF*, Application Note to SpectreRF.

Chaki, S., Aono, S., Andoh, N., *et al.* (1995). Experimental study on spiral inductors. *IEEE Microwave Symposium*, 753–756.

Chang, J., Abidi, A. and Gaitan, M. (1993). Large suspended inductors on silicon and their use in a 2 μm CMOS RF amplifier. *IEEE Electron Device Letters*, **14**, 246–248, May.

Craninckx, J. and Steyaert, M. (1995). Low-noise voltage-controlled oscillators using enhanced LC tanks. *IEEE CAS-II*, **42**, 794–804, Dec.

Craninckx, J. and Steyaert, M. (1995b). A 1.8 GHz CMOS low phase noise voltage-controlled oscillator with prescaler. *IEEE Journal of Solid-state Circuits*, **30**, 1474–82, Dec.

Craninckx, J. (1996). A 1.75 GHz/3 V dual-modulus divide-by-128/129 prescaler in 0.7 μm CMOS. *IEEE Journal of Solid-state Circuits*, **31**, 890–97, July.

Craninckx, J. and Steyaert, M. (1998). A fully integrated CMOS DCS-1800 frequency synthesizer. *IEEE Journal of Solid-state Circuits*, **33**, 2054–65, Dec.

Dehng, G., Yang, C., Hsu, J. and Liu, S. (2000). A 900 MHz 1 V CMOS frequency synthesizer. *IEEE Journal of Solid-state Circuits*, **35**, 1211–14, Aug.

Egan, W. F. (2000). *Frequency Synthesis by Phase Lock*, John Wiley & Sons.

ETSI. (1996). *Digital Cellular Telecommunications System (Phase 2+): Radio Transmission and Reception (GSM 5.05)*, European Telecommunications Standards Institute.

Gardner, F. M. (1980). Charge pump phase-lock loops. *IEEE Transactions on Communications*, **28**, 1849–58, Nov.

Goldberg, B.-G. (1996). *Digital Techniques in Frequency Synthesis*, McGraw-Hill.

Gray, P. and Meyer, R. (1992). *Analysis and Design of Analog Integrated Circuits*, Wiley.

Grebene, A. (1984). *Bipolar and MOS Analog Integrated Circuit Design*, Wiley.

Greenhouse, H. (1974). Design of planar rectangular microelectronic inductors. *IEEE Transactions on Parts, Hybrids, and Packaging*, 101–109, June.

Gutierrez, G., Kong, S. and Coy, B. (1998). 2.488 Gb/s silicon bipolar clock and data recovery IC for SONET (OC-48). *IEEE Proceedings of Custom Integrated Circuits Conference*, 575–8, May.

Hajimiri, A., Limotyrakis, S. and Lee, T. H. (1998). Phase noise in multi-gigahertz CMOS ring oscillators. *IEEE Proceedings of Custom Integrated Circuits Conference*, 49–52.

Hajimiri, A. and Lee, T. H. (1998b). A general theory of phase noise in electrical oscillators. *IEEE Journal of Solid-state Circuits*, **33**, 179–94, Feb.

Hajimiri, A., Limotyrakis, S. and Lee, T. H. (1999). Jitter and phase noise in ring oscillators. *IEEE Journal of Solid-state Circuits*, **34**, 790–804, June.

Hegazi, E., Sjoland, H. and Abidi, A. A. (2001). A filtering technique to lower LC oscillator phase noise. *IEEE Journal of Solid-state Circuits*, **36**, 1921–30, Dec.

Herzel, F., Erzgraber, H. and Ilkov, N. (2000). A new approach to fully integrated CMOS LC oscillators with a very large tuning range. *IEEE Proceedings of Custom Integrated Circuits Conference*, 573–6, May.

Herzel, F. and Razavi, B. (1999). A study of oscillator jitter due to supply and substrate noise. *IEEE Transactions on Circuits and Systems II*, **46**, 56–62, Jan.

Kan, T., Leung, G. and Luong, H. (2002). A 2 V 1.8 GHz fully integrated CMOS dual-loop frequency synthesizer. *IEEE Journal of Solid-state Circuits*, **37**, 1012–20, Aug.

Kim, S., Lee, K., Moon, Y., et al. (1997). A 960 Mbps pin interface for skew-tolerant bus using low jitter PLL. *IEEE Journal of Solid-state Circuits*, **32**, 5, 691–700, May.

Kim, J. and Kim, B. (2000). A low phase-noise CMOS LC oscillator with a ring structure. *IEEE International Solid-state Circuits Conference*, 430–31, Feb.

Krishnapura, N. and Kinget, P. (2000). A 5.3 GHz programmable divider for HiPerLAN in 0.25 μm CMOS. *IEEE Journal of Solid-state Circuits*, **35**, 1019–24, July.

Kundert, K. (1995). *The Designer's Guide to SPICE and Spectre*, Kluwer Academic Publishers.

Larsson, P. (1996). High-speed architecture for a programmable frequency divider and a dual-modulus prescaler. *IEEE Journal of Solid-state Circuits*, **31**, 744–8, May.

Larsson, P. (2001). An offset-cancelled CMOS clock-recovery/demux with a half-rate linear phase detector for 2.5 Gbps optical communication. *IEEE International Solid-state Circuits Conference*, 74–5.

Lee, S.-J., Kim, B. and Lee, K. (1997). A fully integrated low-noise 1 GHz frequency synthesizer design for mobile communication application. *IEEE Journal of Solid-state Circuits*, **32**, 760–65, May.

Lee, S.-J., Kim, B. and Lee, K. (1997b). A novel high-speed ring oscillator for multiphase clock generation using negative skewed delay scheme. *IEEE Journal of Solid-state Circuits*, **32**, 289–91, Feb.

Lee, T. (1998). *The Design of CMOS Radio-frequency Integrated Circuits*, Cambridge University Press.

Lee, T., Samavati, H. and Rategh, H. (2002). 5 GHz CMOS wireless LANs. *IEEE Transactions on Microwave Theory and Technique*, **50**, 268–80, Jan.

Leeson, D. B. (1966). A simple model of feedback oscillator noise spectrum. *Proceedings of the IEEE*, **54**, 329–30, Feb.

Lehmann, T. and Cassia, M. (2001). 1 V power supply CMOS cascode amplifier. *IEEE Journal of Solid-state Circuits*, **36**, 1082–86, July.

Leung, G. C. T. and Luong, H. C. (2003). A 1 V 13 mW 2.5 GHz double-rate phase-locked loop with phase alignment for zero delay. *European Solid-state Circuits Conference (ESSCIRC)*, 109–112, Sept.

Leung, G. C. T. and Luong, H. C. (2003b). 1 V 5.2 GHz 27.5 mW fully-integrated CMOS WLAN synthesizer. *European Solid-state Circuits Conference (ESSCIRC)*, 103–6, Sept.

Liu, T. (1999). A 6.5 GHz monolithic CMOS voltage-controlled oscillator. *IEEE International Solid-state Circuits Conference*, pp. 404–5, Feb.

Lo, C. W. and Luong, H. C. (1999). 2 V 900 MHz quadrature coupled LC oscillators with improved amplitude and phase matchings. *Proceedings of the 1999 IEEE ISCAS*, **2**, 585–8, June.

Lo, C. W. and Luong, H. C. (2002). A 1.5 V 900 MHz monolithic CMOS fast-switching frequency synthesizer for wireless applications. *IEEE Journal of Solid-state Circuits*, **37**, 4, 459–70, April.

Long, J. and Copeland, M. (1997). The modeling, characterization, and design of monolithic inductors for silicon RF ICs. *IEEE Journal of Solid-state Circuits*, **32**, 357–69, Mar.

Mansuri, M. and Yang, C.-K. K. (2002). Jitter optimization based on phase-locked loop design parameters. *IEEE Journal of Solid-state Circuits*, **37**, 1375–82, Nov.

Matsuya, Y., Uchimura, K., Iwata, A., *et al.* (1987). *IEEE Journal of Solid-state Circuits*, **22**, 921–29, Dec.

Merrill, R. B., Lee, T. W., You, H., Rasmussen, R. and Moberly, L. A. (1995). Optimization of high Q integrated inductors for multi-level metal CMOS. *Proceedings of International Electronic Device Meeting 1995*, 983–6.

Mijuskovic, D., Bayer, M., Chomicz, T., *et al.* (1994). Cell-based fully integrated CMOS frequency synthesizers. *IEEE Journal of Solid-state Circuits*, **29**, 271–9, Mar.

Mohan, S. S., del Mar Hershenson, M., Boyd, S. P. and Lee, T. H. (1999). Simple accurate expressions for planar spiral inductances. *IEEE Journal of Solid-state Circuits*, **34**, 1419–24, Oct.

Momtaz, A., Cao, J., Caresosa, M., *et al.* (2001). A fully integrated SONET OC-48 transceiver in standard CMOS. *IEEE Journal of Solid-state Circuits*, **36**, 1964–73, Dec.

Nguyen, N. M. and Meyer, R. G. (1992). Start-up and frequency stability in high-frequency oscillators. *IEEE Journal of Solid-state Circuits*, **27**, 810–20, May.

Niknejad, A., Gharpurey, R. and Meyer, R. (1998). Numerically stable Green function for modeling and analysis of substrate coupling in integrated circuits. *IEEE Transactions on Computer-aided Design of Integrated Circuits and Systems*, **17**, 305–315, Apr.

Niknejad, A. and Meyer, R. G. (1998b). Analysis, design, and optimization of spiral inductors and transformers for Si RF ICs. *IEEE Journal of Solid-state Circuits*, **33**, 1470–81, Oct.

Pache, D., Fournier, J., Billiot, G. and Senn, P. (1995). An improved 3 V 2 GHz BiCMOS image reject mixer IC. *IEEE Proceedings of Custom Integrated Circuits Conference*, 95–8, May.

Parker, J. F., and Ray, D. (1998). A 1.6 GHz CMOS PLL with on-chip loop filter. *IEEE Journal of Solid-state Circuits*, **33**, 337–43, Mar.

Perrott, M. H., Tewksbury III, T. L. and Sodini, C. G. (1997). A 27 mW CMOS fractional-*N* synthesizer using digital compensation for 2.5 Mbps GFSK modulation. *IEEE Journal of Solid-state Circuits*, 2048–60, Dec.

Poore, R. (2001). *Phase Noise and Jitter*, Agilent Technologies.

Porret, A., Melly, T., Enz, C. and Vittoz, A. (2000). Design of high-Q varactors for low-power wireless applications using a standard CMOS process. *IEEE Journal of Solid-state Circuits*, **35**, 337–45, Mar.

Pottbacker, A. and Langmann, U. (1994). An 8 GHz silicon bipolar clock recovery and data regenerator IC. *IEEE Journal of Solid-state Circuits*, **29**, 1572–8, Dec.

Rategh, H. and Lee, T. (1999). Superharmonic injection locked oscillators as low power frequency dividers. *IEEE Journal of Solid-state Circuits*, **34**, 813–821, June.

Razavi, B. and Sung, J. (1994). A 6 GHz 60 mW BiCMOS phase-locked loop with 2 V supply. *IEEE International Solid-state Circuits Conference*, 114–15, Feb.

Ravazi, B. (1996). *Monolithic Phase-Locked Loops and Clock Recovery Circuits*, IEEE Press.

Razavi, B. (1996b). A study of phase noise in CMOS oscillators. *IEEE Journal of Solid-state Circuits*, **31**, 221–343, March.

Razavi, B. (1997). Challenges in the design of frequency synthesizers for wireless applications. *IEEE Proceedings on Custom Integrated Circuits*, 395–402.

Razavi, B. (1998). *RF Microelectronics*, Prentice-Hall, Inc.

Riley, T. A. D., Copeland, M. A. and Kwasniewski, T. A. (1993). Delta–sigma modulation in fractional-*N* frequency synthesis. *IEEE Journal of Solid-state Circuits*, **28**, 553–9, May.

Rofougaran, A., Rael, J., Rofougaran, M. and Abidi, A. (1996). A 900 MHz CMOS LC oscillator with quadrature outputs. *IEEE International Solid-state Circuits Conference*, 392–3, Feb.

Rogenmoser, R., Huang, Q. and Piazza, F. (1994). 1.57 GHz asynchronous and 1.4 GHz dual-modulus 1.2 μm CMOS prescalers. *IEEE Custom Integrated Circuits Conference*, 387–90, May.

Shu, K., Sanchez-Sinencio, E., Silva-Martinez, J. and Embabi, S. (2003). A 2.4 GHz monolithic fractional-*N* frequency synthesizer with robust phase-switching prescaler and loop capacitance multiplier. *IEEE Journal of Solid-state Circuits*, **38**, 866–74, June.

Sidiropoulos, S., Liu, D., Kim, J., Wei, G. and Horowitz, M. (2000). Adaptive bandwidth DLLs and PLLs using regulated supply CMOS buffers. *Symposium on VLSI Circuits*, 124–7, June.

Steyaert, M. and Craninckx, J. (1994). 1.1 GHz oscillator using bondwire inductance. *IEEE Electronics Letters*, **30**, 244–5, Feb.

Su, D., *et al.* (2002). A 5 GHz CMOS transceiver for IEEE 802.11a wireless LAN. *IEEE Journal of Solid-state Circuits*, **37**, 1688–94, Dec.

Tadjpour, S., Cijvat, E., Hegazi, E. and Abidi, A. (2001). 900 MHz dual-conversion low-IF GSM receiver in 0.35 μm CMOS. *IEEE Journal of Solid-state Circuits*, **36**, 1992–2002, Dec.

Vaucher, C. (2000). An adaptive PLL tuning system architecture combining high spectral purity and fast settling time. *IEEE Journal of Solid-state Circuits*, **35**, 490–502, Apr.

Vaucher, C., Ferencic, I., Locher, M. *et al.* (2000b). A family of low-power truly modular programmable dividers in standard 0.35 μm CMOS technology. *IEEE Journal of Solid-state Circuits*, **35**, 1039–45, July.

Wang, H. M. (2000). A 1.8 V 3 mW 16.8 GHz frequency divider in 0.25 μm CMOS. *IEEE ISSC Digest of Technical Papers*, 196–7, Feb.

Wang, H. M. (2001). A 50 GHz VCO in 0.25 μm CMOS. *IEEE ISSC Digest of Technical Papers*, 372–3, Feb.

Wong, J., Cheung, V. and Luong, H. C. (2002). A 1 V 2.5 mW 5.2 GHz frequency divider in a 0.35 um CMOS process. *Symposium on VLSI Circuits*, 190–93, June.

Yamagishi, A., Ishikawa, M., Tsukahara, T. and Date, S. (1998). A 2 V, 2 GHz low-power direct digital frequency synthesizer chip-set for wireless communication. *IEEE Journal of Solid-state Circuits*, **33**, 210–17, Feb.

Yan, W. and Luong, H. C. (2001). A 900 MHz low-phase-noise voltage-controlled ring oscillator. *IEEE Transactions on Circuits and Systems II: Analog and Digital Signal Processing*, **48**, 2, 216–21, Feb.

Yan, W. and Luong, H. C. (2001b). A 2 V 900 MHz monolithic CMOS dual-loop frequency synthesizer for GSM wireless receivers. *IEEE Journal of Solid-state Circuits*, **36**, 2, 204–16, Feb.

Yoshizawa, H., Taniguchi, K. and Nakashi, K. (1998). An implementation technique of dynamic CMOS circuit applicable to asynchronous/synchronous logic. *1998 IEEE ISCAS*, 145–8, June.

Young, I. A., Greason, J. K., Smith, J. E. and Wong, K. L. (1992). A PLL clock generator with 5 to 110 MHz lock range for microprocessors. *IEEE Journal of Solid-state Circuits*, **27**, 1599–1607, Nov.

Yuan, J. and Svensson, C. (1989). High-speed CMOS circuit technique. *IEEE Journal of Solid-state Circuits*, **24**, 62–70, Feb.

Yue, C. and Wong, S. (2000). Physical modeling of spiral inductors on silicon. *IEEE Transactions on Electron Devices*, **47**, 560–8, Mar.

Zhang, P. *et al.* (2003). A direct conversion CMOS transceiver for IEEE 802.11a WLANs. *International Solid-state Circuits Conference*, 354–5, Feb.

Zheng, J., Hahm, Y.-C., Tripathi, V. and Weisshaar, A. (2000). CAD-oriented equivalent-circuit modeling of on-chip interconnects on lossy silicon substrate. *IEEE Transactions on Microwave Theory and Techniques*, **48**, 1443–51, Sept.

Index